Charismatic Cows and Beefcake Bulls

Charismatic Cows and Beefcake Bulls

SONIA KURTA

Old Pond Publishing

First published 2007

ISBN 978-1-905523-76-4

A catalogue record for this book is available from the British Library

Published by
Old Pond Publishing
Dencora Business Centre
36 White House Road
Ipswich
IP1 5LT
United Kingdom

www.oldpond.com

Cover design by Liz Whatling
Typeset by Galleon Typesetting, Ipswich
Printed and bound in Great Britain by Biddles Ltd, King's Lynn

Contents

Acknowledgements

I am indebted to those who searched their archives to provide photographs for my book: Diana Wall of the Minchinhampton Parish Office, all those at the Cornish Studies Library in Redruth and to Robert Cook at the Courtney Library Photographic Collection in Truro. The front cover photograph is courtesy of the Museum of English Rural Life, Reading. Back cover photograph: Heather Jarrold.

Chapter 1

How It Started

'GOOD MORNING, gal,' says the cowman. 'Life's one continual round of pleasure, ain't it!'

It's 5.30 on a cold, wet, winter's morning. The only thing relieving the total blackness of the cowshed is a small patch of light from the hurricane lantern.

We fetch our buckets and stools and start milking. Take the bucket in the right hand and the stool in the left, swing the first under the cow and the second under your bottom. Get them in the wrong hands and you will end up sitting in the bucket which, at that time of day, is even easier to do than normal.

On this early morning in 1943 I was too sleepy to chatter to the cowman. Instead I began wondering why on earth I was working at this hour every day – Saturdays, Sundays, Christmas Days, all included. This was not my first farm and not the first time I had thought myself mad to be doing what I was doing. I began to recall how innocently it had all started. . . .

When I was a child my grandparents took early retirement in their fifties. They bought about six acres of land near Tring in Hertfordshire where, in the days of easier planning permission, they built a bungalow.

My mother had an old-fashioned wooden gipsy caravan built on Grannie's land. Mother's idea of touring with a horse between the shafts was scotched because my mother said that the caravan's iron-rimmed wheels were not allowed on the road. It just sat there and we spent our holidays in it.

Granny and Grandad were complete opposites. Grandad was a tall, thin, sinewy man. I don't remember ever seeing him smile, let alone laugh. He was not miserable, just of a serious turn of mind.

Granny was a well-rounded person who would laugh at the comedians on the wireless until the veins stood out on her forehead. She handled all the money and gave Grandad just enough for his ounce of baccy and a few sweets. She was also a very good gardener and people would stop as they passed just to admire her front garden. One gentleman told her that his professional gardener could not get things to grow as well as she could.

Granny and Grandad were self-sufficient, growing all their own vegetables and keeping a lot of hens. I can't think why they had so many. When my uncle made a swing for me and the local children, we taught some of the hens to ride on our laps. They used to flap their wings to keep their balance while we swung to and fro.

When my grandparents bought the land they promptly sold three acres on which a children's hostel was built. This was mainly for the children of civil servants and colonial officers who could not go and join their parents in the holidays.

The children's hostel managed to break a large amount of thick, white crockery. Grandad would collect it, then smash it with a hammer and a special stone for the chickens who gathered round to gobble it up. Once we were old enough not to hammer our fingers, or at least not to scream too loudly when we did, we were allowed to undertake this task.

Grandad became a bit of a goat man and had two young goats, a brown Toggenburg called Sally and a white Saanen called Betty, who knew that we were young and they could play with us.

They would rear up on their back legs and come crashing down three or four feet away, then walk over gently, lowering their heads to play push. They seemed to enjoy it as much as we did.

Grandad acquired his first goat because she was being mistreated. She was called Nanny and she was a horror. If she got free, look out everybody!

She once got loose when my mother was coming back from Grannie's bungalow with a jug of milk.

Nanny took off after my mother who did not have time to get into the caravan and had to do quite a few circuits around it before she got far enough ahead of the goat to be able to get up the steps.

Another time when my mother was away from the caravan fetching the milk, two goat kids skipped up the steps and danced all over the table laid for tea. I arrived to hear her exclaiming, 'They've put their feet in the butter and there's jam all over the bed!'

I used to tease Nanny, and I suppose my Grandad too, by imitating a goat's bleat from outside her shed so that she would stick her head out to answer the stranger. The only person she liked was my Grandad, so he would come running out of the bungalow, thinking that something was wrong.

Grandad's next acquisition was a billy-goat called Rastus who had been living with a greengrocer who probably thought that Rastus would be useful to mop up the odds and ends that he didn't use. Unfortunately, Rastus didn't take kindly to being a mopper upper, so he used to escape. He would go out to the front where the greengrocer had his displays and cause havoc. After he had done this a few times the greengrocer got fed up and, hearing that my Grandad had a goat or two, asked if he would take Rastus.

Grandad had a routine in the mornings which we would gather round to watch with glee. He had never been a heavy man and by this time was quite elderly and frail. Rastus was a big, strong billy-goat. Grandad would go into Rastus's shed and shut the door behind him. He put a chain onto Rastus's collar and a stake ready through the ring at the end, then, with a mallet in the other hand, he would open the door. Rastus would burst forth and charge off with Grandad flying behind him.

As soon as the goat stopped to grab a tasty morsel Grandad would hastily start hammering the stake in. If the ground was soft he would often succeed but if the ground was hard Rastus would take off before he had a chance. Mallet and stake went one way, Grandad went the other, and Rastus got into Grannie's prized flowers before he was caught.

For a child, farm life was great fun; the animals were interesting and, of course, we had all the fresh vegetables that Grandad was growing. I was always happy to help – have you noticed how children are keen to help until they get to the age when they could be useful?

On one occasion I was trotting about with my barrow when two ladies called out, 'One day you will have to be a Land Girl.' The idea sounded much more interesting than what they were teaching us at school, so the comment set a seed in my mind.

When we were not on holiday we lived near, but not in, London and I travelled to school by train. Just before war broke out my mother thought that the Germans would probably be aiming at the railway lines as they had in the First World War and I would be safer not travelling to school. She unofficially evacuated me to Gloucestershire, not in the vale of Gloucester but high up on the Cotswolds. These hills consist of a huge slab of limestone. We could cycle down to school in Stroud in less than half an hour but it took nearly two hours to get home pushing our bikes all the way from Thrupp at the bottom to Rodborough at the top. At the top it was so flat that an airfield had been built there and Minchinhampton Common had had to be covered in ditches and concrete blocks to discourage the Germans from trying to land on it.

I was sent to live with two retired school teachers, one of whom had been my head mistress at a little private school. This was very unfair, both to me, a thirteen-year-old who regarded them as terrifying creatures, and to them. They must have been looking forward to their retirement and have had enough of kids, especially one like me, whose sole aim in life was to see just over the next hill and around the next corner.

After I had been there for a few months I went home because my father was dying. In my absence my unwilling hosts moved a friend of theirs, another retired teacher, into my room so that when the time came for me to go back, there was no space.

In the meantime I had made friends with a girl whose grandparents owned a small dairy farm which was also in the Cotswolds. She and her brother were living there because their parents were working in Birmingham, which was not the place to be in 1940 if you could avoid it. I went to live with them.

A few months after my father died the chemist businesses which he had left were bombed. He had had two premises in London. One was totally destroyed and the other sufficiently damaged to put an end to its use.

This left my mother in a very precarious position so when my friend's family offered to have me without charging very much for my keep my mother was willing to accept. This was providing I helped on the farm at weekends, the holidays and, as it turned out, before and after school.

On this farm I came across one of the most memorable cows that I have met. She was a Guernsey called Brownie.

The herd is led by a dominant cow and the others have their recognised places under her. She is not necessarily the biggest or the strongest but has a characteristic that I can only describe as charismatic. Brownie was exactly that.

On 13th May, commoners were entitled to turn their cattle out onto Minchinhampton Common. Before this began there was a day when all cattle had to be presented for marking before they were allowed to graze. We rounded up the animals and drove them to a corral in the centre of the common while trying to keep them from getting mixed up with the other beasts. People were on foot and mounted on bikes or horses, shouting and frantically waving arms and sticks. It was a great day which, as youngsters, we really enjoyed.

To benefit from the free grass we would turn the cows out onto the common after morning milking and bring them home after school for evening milking.

One morning a week everyone would put their dust-bins out and Brownie would charge ahead of the rest, udder swinging, and proceed to knock the lids off the bins to look for goodies inside. In the process of her search she pulled out all the contents, leaving rubbish strewn across the street. We couldn't leave the other cows to go after her because they would have scattered.

Since she was the leader, Brownie was always prepared to challenge any cow who she didn't recognise as being part of her herd. Unfortunately this included herself and, on seeing her reflection in the plate glass windows of the shops that we passed, she would lower her head and prepare to charge. We were always just in time to stop her – but talk about a bull in a china shop; she would have been a cow in a clothes shop and draped in all the latest fashions.

Also among the entertaining animals in the Cotswolds

was a pig which showed a considerable ability for working things out. Any chicken making the mistake of hopping into a pig sty in the hope of finding something to eat has almost certainly signed its death warrant, as the pigs will grab it and eat it. One old sow had taken this a step further. She had a scheme for getting this delicacy for herself.

Most of the time she was running free and the poultry spent the night in houses of about thirty to fifty birds. When they had gone to bed through the pop hole Sal Slap-Cabbage (all sows seemed to be called this), would go round the back of one of the houses, which stood a few inches off the ground, put her snout underneath, and shake the house up and down as hard as she could. The hens would panic and start to rush out though the pop hole, whereupon the gourmand would charge round and grab a nice fat bird.

While living here I learned a lesson which was to prove valuable for the rest of my time working on farms. It is generally thought that cows are not as dangerous as bulls but a cow with a calf is liable to attack anyone, especially people they don't know.

When my friend Betty and I were about fourteen or fifteen years old, a new cow called Fillpail was purchased with calf at foot. When we were trying to get her into the cow shed we were very careful not to get between her and the calf. However, perhaps she was not used to children's voices because she attacked Betty who would have been severely gored if she had not had the presence of mind to hang onto Fillpail's front leg. Fortunately I managed to grab a shovel and drive the cow away. Betty still has trouble with her back to this day as a result of this incident.

In those days cows were not dehorned. I saw many people injured as a result of this, including, several years later on a different farm, a relief stockwoman who lost an eye. I was relieved when cows began to be dehorned so as to do less damage to each other.

I remained at school until I was sixteen which must have been a big struggle for my mother because she had to pay fees, but I got more and more bored because I couldn't do the subjects that I wanted.

In those days you had to do either sciences or arts. I was very interested in biology but also liked English and it seemed to me that they didn't realise that if you studied a science, you might also want to write about it. Finally, since I had dropped mathematics as soon as I was able, but could not do science without it, I was made to drop biology. I had had enough of school.

My friend Betty had gone off to Cannington agricultural college to study dairying and I should have liked to have gone somewhere similar but my mother could not afford to send me. At the time there were scholarships available for the sons and daughters of farm workers or for bona fide farm workers but for the latter you had to have been doing farm work for at least five years, so I continued working on farms in various parts of the country.

What follows is an account of my experiences and some of the people and animals I met during the next eleven years.

Chapter 2

Farms and Farmers

I FINALLY entered full-time farm work in 1942 on Bubblewell Farm, the little dairy holding where I had been living for two years.

The farm was owned and run by my friend's grandmother. She was in her early sixties, nut brown and sinewy, and I am sure she must have been very beautiful when she was younger. The problem was that she would have had a hard time organising a cat fight; although she was full of energy it was all totally misplaced.

It was a small farm which before the war had been kept mostly in grass for grazing and to make some hay. Other feed was bought in. I am sure that it could have been a very good little farm in the right hands but when the farmer's son (my friend's uncle) had to give up farming through illness, I was left totally in charge. Although I realised what needed doing, I had neither the skill nor the knowledge to carry it out.

I am sure that with a little effort we could have sold

much more produce since we already had direct contact with the public through the milk round. There was an old orchard producing wonderful plums, apples and pears but, despite a half-hearted effort to gather them, they were generally left to rot. During the war everything made money. Eggs were probably rationed by then so I am sure it would also have paid to have had more hens.

We had some off-lying land where the young stock used to spend most of their time and one day I noticed that there was very little feed left. I told the boss this, thinking that she would arrange for some to be taken up there. Instead, all she said was, 'Sufficient unto the day is the evil thereof.'

When I finished morning milking I would take the dirty buckets down to the dairy to rinse before I changed to go on the milk round. Often my boss would say, 'Oh, you leave those, I will do them,' but I would return to find them exactly where I had left them. It was essential that they were rinsed out very quickly with cold water otherwise the milk would stick and so I would have to set to with extra energy and clean them. Now what can you do on a farm like that?

It was not a lack of knowledge which was the problem but a complete absence of organising skills, although it would be several years before I was to realise just how much this mattered.

The level of hygiene would have given a modern health inspector a nervous breakdown. There was a cold water hose in the dairy and a tap outside the back door, but no electricity and no means of heating water except a big kettle over an open fire. The cow shed was over a hundred yards from the dairy on top of a bank and we carried the milk down in open buckets. Whether it got more dirt in it on a dry, dusty day or when we were slipping about in mud is a moot point.

By this time milk was rationed and we were delivering to people who had probably never had any before. Our hygiene may have been basic but we made some effort towards cleanliness. Some of our customers did not make the same effort and jugs would be covered in thick layers of sour milk. I got strange looks if I refused to fill them until they had been cleaned. A few of our customers were completely the opposite. We normally measured milk from the buckets with a dipper but they believed that milk which came in bottles would be cleaner. For them we kept a few bottles in the van which we filled from the bucket with the dipper anyway. It kept them happy.

Dairy cows give far too much milk for one calf and at the time we experimented with several ways to collect the extra. The most crude way was to fight the calf for who could get the most; you on one side of the cow and the calf on the other. The cow would let the milk down well by this method but it was not very pleasant to have to get hold of the slimy teat that the calf had been working on. Heavens knows what went into the bucket of milk.

A more sophisticated method was for one cow to suckle a number of calves depending on how much milk she was giving. Providing this was done skilfully it produced better calves than the more common method of giving milk substitute in a bucket. Once the calves realised that they had to drink with their heads down they tended to gulp, swallow air and become pot-bellied.

The most hazardous part of milking was the kickers. One old boy who came in a few times a week to help out asked me if any of the cows kicked. When I pointed out the few that did, he told me that I would have to milk them since my bones would mend faster than his if they got broken. You can't expect everyone to be a gentleman.

Kickers were normally dealt with by putting a strap or piece of rope in a figure of eight just above their hocks. A roan cow called Peaches simply jumped up and kicked out with both feet together. One day she jumped right over me and I was left holding a bucket between my knees with no cow to milk.

We fixed her by tying a pole from the manger to the back of the standing which held her firmly in place. She decided that she was beaten and would fall asleep leaning against it – but I still checked the ropes often to make sure that she couldn't get her own back by falling on me.

There was a tale going around about a man who had been concreting his yard and had had some wet cement left over when he started milking. One of the cows lashed out and kicked the bucket of milk over so he replaced it with a bucket of concrete and said, 'Now kick that.' She did. I don't know how long it took her to put that foot to the ground again.

When the war came there were orders to plough up the flatter land. Since the farm did not have the equipment, this was done by a contractor and we were left to deal with the crop. Although a hay rick was nothing new, none of us had ever built a corn stack so a retired farm worker was called in to do the job. I worked on the stack passing him the sheaves. After a bit he said to me, 'You're quiet for a modern youngster' (nothing ever changes, does it). 'That's the way it should be,' he continued. 'Always remember to keep your mouth shut and your eyes, ears and bowels open, and you won't go far wrong.' I have always tried to follow his advice, although the first bit gets harder the older I get.

Since we had no grinder, the corn we produced was taken to the local mill in Avening, about two or three miles away, and we collected it a few bags at a time.

This was one of my favourite jobs and the mare enjoyed it as much as I did, trotting merrily along. The mill was run by electricity but in years gone by it had been water powered and there was still a pond at the side. The track between the pond and the mill was just wide enough for one vehicle.

One spring day when I drove onto the track the miller's wife ran out to meet me. She was holding a wooden rake with two teeth missing and she told me to wait where I was until she called. I was puzzled by this until I turned the corner to see a most indignant male swan with his head held down by the hay rake. The female swan was sitting on her nest on the other side of the pond and the miller's wife told me that the cob would attack anything that came past. It's a good job she saw me coming because if he had attacked the mare she would have panicked and both of us and the trolley could have ended up in the pond.

I enjoyed doing jobs with the horse, especially chain harrowing in the spring and sweeping the hay to the rick after harvest.

The chain harrow pulled out some of the old grass to make the new grass grow better and the chains made a lovely tinkling noise. Everyone tried to drive in dead straight lines and from a distance the patterns would show how successful you had been.

The way we made hay was very old fashioned. A

neighbour would cut the grass for us and then we might borrow a hay turner. Sometimes we turned it by hand with wooden rakes instead and if we stroked the hay in the right direction it would turn over in continuous waves.

When it was dry enough we built it into cocks and it was these cocks which I swept to the rick. The sweep was made entirely of wood with two curved handles and long prongs. To get the hay onto the sweep you had to lift the handles ever so slightly otherwise the sweep would ride right over the hay cock. When you got to where you wanted to dump the hay you had to lift the handles up so the prongs dug into the ground, the sweep somersaulted over the hay and came up into position ready to set off again.

The farm had a few geese which I always found great fun. Whenever I walked past them the gander would follow me. I would ignore him until he was just about to give me a sharp nip then turn and confront him. He immediately pretended that he didn't have the slightest interest in me but had seen something that he had to investigate in the corner. When he returned to the other geese they would gather around and cackle. He was obviously telling them that he had seen me off.

The geese shared a house with some hens who had a pop hole to come and go while the geese used the door. One spring a goose made her nest by a wood pile and my friend's grandfather was watching when the gander apparently decided that the eggs were unsafe. He rolled an egg to the poultry house door but try as he might he could not get it over the step. Instead, he rolled it up the ramp to the pop hole and inside. He repeated this until he had moved all the eggs to safety. It is a good job that goose eggs have strong shells.

Although they can be very intelligent, geese can also be very stupid. If goslings are left with the adults at night they will get trampled. It is wise to separate them. The process leads to a lot of commotion: hissing, spitting, and charging with outstretched wings at the person who they think is trying to kidnap, and possibly murder, their babies.

As youngsters we thought this was great sport. It took at least two of us, one to catch the goslings and one to deal with the angry geese. The only way to deal with the gander was to catch him by the neck and stuff him between your knees to stop him flapping his wings.

Early in the 1940s artificial insemination was being developed but it was not widely used. The accepted practice for farms with only a few cows was to take them to a neighbouring farm which had a bull, and this is just what we did even though sharing a bull ran a real risk of spreading disease between herds.

When we took our cows to the neighbours their son would be waiting on horseback since it was not thought safe to go near the bull on foot. I took a great interest in this young man because he was one of the best horsemen that I ever came across. I once saw him gallop full tilt alongside a wall, leap onto the wall, grab an apple off a tree and then leap back into the saddle. Mind you, I don't think the cows appreciated it. When I was milking I would often look across the valley to see him jumping over walls and chasing the poor cows at a gallop.

I met him a few years later at a dance and was pleased to find that he was as clumsy as a cow with a musket, as the saying was, and that there was something which I could do better than him (although not much).

The Cotswolds were once famous for their own breed of sheep which had provided wealth through the wool trade. But by the time I was working there small dairy farmers could not spare grazing for sheep as they were not as profitable. Bubblewell Farm didn't have any but there were a few on the farm next door which sometimes got onto our fields.

On one of these occasions when we asked the farmer to come and remove his sheep his reply was: 'They b'aint mine, they be our Alan's. I can't abide ship, they be always goin baa – you veed em one minute then come back vive minutes later and they still be gwoing baa.'

I am sure I would have learned a lot more from my

time on Bubblewell Farm if Betty's uncle Jim had been in good health. He was not yet forty but had been suffering from a severe form of arthritis for several years. That winter he caught 'flu and when the doctor examined him he was told he had to give up farm work immediately. It was a shame. Despite his problems he was always cheerful and had a fund of outrageously colourful stories and sayings. Working with him was one big laugh.

One day a customer stopped me on the milk round and told me that her daughter was looking for a job on a farm. It was 1943 and all unmarried women between eighteen and forty had to join the forces or do work of national importance. I went back to the farm pleased because I thought that even though she wouldn't know any more than me, at least she would be able to help. When I suggested it to the elderly owner she said, 'Oh no, you can manage on your own.'

Of course, I couldn't manage on my own and was working eighty hours a week for 8 shillings. I decided to leave.

Chapter 3

The Three Meanies

MY next farm job was in Surrey. This part of the country was in complete contrast to the Cotswolds with heavy clay, hence the name, Claygate. The farm was a mixed farm of about 150 acres and although it was unsuitable for sheep it gave a high yield of potatoes which earned the farmer his local nickname 'The Potato King'. There was also a dairy herd where I worked as the cowman's assistant. Whether the farm had always been this size I don't know but the farmhouse was a large, imposing building with a superior cottage attached. It was probably intended for the farm manager or bailiff but was now occupied by an army officer's widow.

The farm was owned by a bachelor farmer and his two spinster sisters who turned out to be the meanest, most suspicious minded and generally miserable people I have ever worked for. The farmer was in his sixties and one of those tall gangly people who look as though their arms and legs are not properly screwed on. He was very freckly and probably had ginger hair in his younger days, so since his complexion was bright red, he must have clashed with himself most of the time.

I was amazed by how superior the farm family and the widow considered themselves even though they were entirely dependent upon their workers.

The widow had a friendly housekeeper who would often invite me in for coffee and her delicious cakes. When her sister was taken ill and she had to visit her for a few days, her mistress was totally incapable of cooking herself even the simplest meal and had to live on bread and cheese.

I quickly realised that the farmer was just as helpless. There was a tractor which we used for heavier work as well as a car in the garage, laid up because of the war; he could drive neither.

One day the cowman asked him to come and look at one of the cows because the cowman thought that there was something seriously wrong with its udder and the vet should be called. We then realised that he had not even the faintest idea how to get milk out of a cow. When I told my mother about this she said, 'Well, he is a gentleman farmer.' I firmly buttoned my lip. She would not have appreciated my comments.

I never had a lot of respect for these so-called gentlemen farmers. A few years later when I had joined the Land Army I was interviewed by a potential employer and asked if I would be able to do some tractor driving. He said I certainly could and then commented, 'I was quite surprised to find out how difficult it was when I went out to help the men last harvest time.' The Land Army authorities could not understand why I refused to take the job.

As I have said, the most outstanding feature of this farm family was their meanness. When I arrived at the farm one of the men told me that the farmer would follow the harvest wagons on his sit-up-and-beg bicycle and if he saw an ear of corn fall he would pick it up and put it in his bicycle basket. I could not believe that anyone could be so petty and assumed I was having my leg pulled, but I actually saw him do it on more than one occasion. I took great delight in getting my own back at every opportunity.

Across the yard from the family's back door was one of the big barns where hay and straw were kept, usually with a few hens poking about. I would put my foot under one and gently lift her into the air which gave her such a fright that she would start cackling. One of the sisters would come running out thinking she'd laid an egg and say to me, 'Did you see where that hen came from?' I would point at the hayloft and reply 'Yes, she flew down from there.' I knew they wouldn't ask me to go up and look for the egg because they would think that I would say I couldn't find it and then return later to pinch it. It amused me terrifically to watch them struggle up the ladder to look for a non-existent egg.

There was a lad who worked there who was about the same age as me and we became friends. He kept pet rabbits and while he waited for me after work he would fill a little sack for them with green stuff from the side of the road.

One day I came out and there he was, red in the face, pushing all the green stuff back into the sack. The two old dears were waddling away towards their chicken pens. 'They came along,' he told me, 'and said, "What have you got in that sack? Turn it out."'

I thought, right, we'll feed the rabbits some of their food, and after that I used to grind up a bit of corn for my friend to feed his pets.

The cowman with whom I worked left (a good story to which I shall return), but stayed in his cottage until he found another job. He asked me if I could smuggle him some milk. He had a newspaper delivered to the dairy and every evening I put a big bottle of milk into my pocket then buttoned a jacket over it and waltzed over to his house with the paper under my arm. I often met the sisters on their way back from the poultry pens and with a big grin would most politely wish them a good evening, practically dropping them a curtsy. I was so obviously pleased with myself that it is a wonder they never smelt a rat.

We started milking every morning at 5.30 and by 6.30 the boss would come wandering out with a big jug. Now, evening milk is always richer than the morning milk and because it has been in the churns all night, the cream is sitting on the top. He would dip his jug into the top of the churn and off he would trot with his jugful of cream.

When the old cowman left, after a few days the new man said to me, 'Girl, I'm not having that, he's doing the public out of their rights.' In those days milk was sold just as it came out of the cow, not like nowadays when they take bits out and put bits in so you never get anything like it was originally. The cowman said, 'Your first job, girl, when you get here in the morning, is to give those churns a really good stir.' Oh boy, did I enjoy doing that!

The farmer never said anything but I bet he didn't half get into trouble when he went into the house. I can just imagine what the sisters said: 'What's the matter with the milk?'

On every farm where I worked I came across people who were afraid to stand up for themselves. They would complain about how badly they were treated, but when I asked why they didn't say anything, they would reply, 'If you had ten men waiting to jump into your job and a family to feed, you wouldn't say anything either.' Even during the war when there was no-one else to do the job, the attitude was the same. Not that they were lazy. They worked hard and cared about their job, they just did not want the responsibility of speaking up.

The Three Meanies, as I have come to think of them, taught me the need to stand up for myself.

Many of us had continued to work over the August bank holiday and so were due overtime. One of the men warned me, 'If you don't ask for overtime, you won't get it.' Forewarned is forearmed so when pay day came I knew exactly what I was owed.

Normally I was last in line because we had to tidy up the dairy before going home at Saturday dinner time. When I arrived at the farmhouse there was no-one in sight but as soon as I knocked, all the crafty men came out of the barn where they had been hiding. The older sister who dealt with the wages handed me my money – no pay packets or anything fancy like that, just counted into my hand. I said, 'Excuse me, but I worked extra hours over the bank holiday so you owe me some overtime.' There was no way she could get away without paying me so she had to go and fetch it. I expect she could have killed me. One by one, with big grins on their faces, the men came forward to tell her what she owed them.

Of all the extraordinary characters I met while working in Surrey, the most memorable was the cowman, Frank, a marvellous man of sixty-three. He started work every day at 5.00 and was still enthusiastic even though he had been working with cows since he was fourteen. He told me that when he was younger he would sometimes be so tired that he would fall asleep while he was working.

Frank had an amazing amount of knowledge and gave me a lot of tips. He could make ointment out of willow herb and told me that if a cow lost its appetite through illness I should give it ivy as a pick-me-up. He also taught me how to use the barn machinery safely: 'When you're using the machinery put a cap on, no loose hair, roll your sleeves up tight, no loose clothing. I've seen a man's trousers taken off over his head and it's not a pretty sight.'

I loved the grinding, chaff cutting and mangold pulping because it was done with a belt-driven machine which was much better than doing it by hand as I had in Gloucestershire. This machine was run by an electric dynamo with the belt-driven pulleys going in all directions. They made a lovely slapping noise, especially when they got just a little bit loose. We used to put Stockholm tar on them; it was used for everything that required a sticky or antiseptic substance.

Frank had a son in the Merchant Navy and one Christmas this lad brought Frank back a bottle of whisky which, of course, Frank was not used to drinking. Then on Boxing Day he didn't turn up. I didn't dare go over to the cottage to see where he was because I could have milked at least one more cow in that time and it was always a rush to catch the milk lorry at the best of times. In the end he did turn up and he was rolling around like a sailor.

The cow's food was mixed in a big heap on the barn floor – pulped mangolds, chopped kale, rolled grain and chaff. We shovelled it into tin bushels to go to each cow in the shed. Frank was staggering so much that I was sure he would go head first into the feed so I shovelled frantically to make sure there was always a bushel waiting for him when he came out of the cow shed.

When he came back in the afternoon he had had a drop more whisky and the boss foolishly came out into the yard; I suppose his sisters had turfed him out. Well, Frank, full of Dutch courage, told the boss where to put his job. When he came in and told me what he had done he kept saying, 'It's not your fault, you're a good girl,' and trying to pat me on the head.

I always thought old Frank probably had a heart problem because of the bright patches on his cheeks. Since it was dangerous to smoke around so much hay and straw he used to chew tobacco and it was well known that this caused heart problems. Apart from anything else, he worked too blooming hard. He found a new job with a well-known farm quite quickly but had only been there a few months before I heard that he had dropped down dead.

After Frank left we had a stand-in for a few months, a smart fellow who thought it was clever to be dishonest.

He told me about a farm he had been on which had had foot and mouth disease. Animals with the disease were supposed to be destroyed, not because they couldn't be cured but because they wouldn't thrive afterwards and would increase the risk of an epidemic the longer they were kept. Of course, he had cured them and kept quiet.

On another farm he had had a boss who didn't take very much interest in the herd. There was a neighbouring farm which had some very nice cows and he waited for his chance when they had a new cowman. Our sly stand-in broke down the fence so it looked as if the cows had broken through and then went and mixed them up. He picked out the ones that he wanted since the other cowman was too new to be able to recognise his own cows.

The motivation was simple. Very often cowmen were paid a bonus for how much milk they could produce. This could be a big mistake and lead cowmen to put water in the milk, or antics such as this.

In spite of his know-it-all attitude and assurances that he had worked with cattle all his life, one afternoon this cowman proved that he had no idea of the way to deal with a bull who regarded him as a stranger.

Since the farm had a herd of twenty-four cows it had its own bull, a large Red Lincoln Shorthorn.

When I first arrived Frank told me to make friends with the bull and so I carried him his food, kept his bedding clean and groomed him. He especially liked to have the itchy hay seeds brushed out of his pate.

He looked on me and Frank as friends but there were some people whom he violently disliked and would certainly have hurt if he could have got free. These people included strangers, the boss (sensible animal) and the Land Girl who was the rat catcher's mate. If they came into the cowshed he would roar, roll his eyes and paw at the ground.

One afternoon I opened the door to the cowshed to find the frame filled with the huge head of the bull. I hastily shut the door in his face and retreated to have a think. Since the chain was still around his neck it seemed he had just come unhitched from the standing so I put a strong piece of wire and some pliers in my pocket. I thought I would take a bucket of food and then do a hasty repair job while he was munching.

However, just as I was filling the bucket, in came the new cowman and asked me what I was doing. He said he would fix the bull and disappeared into the shed with a stick.

I couldn't see what was happening. When the

cowman's daughter came along and I told her where her father was, we decided that the only way we could see safely was to go to the windows on the yard where we had to climb on each other's backs in order to see. There was the cowman, perched on the rafters with the bull beneath him looking straight up.

We managed to persuade the bull to come out into the yard and the cowman climbed stiffly down from his precarious position. When his daughter asked what on earth he was doing he explained that he had thought that he would just slip between two of the cows. Well, to his surprise the bull followed, although I don't know how. There was barely enough room for a small human let alone a huge bull.

Having got the bull into the yard we decided to try and put him into one of the loose boxes. Then the boss came out to see what was happening and the bull, who had climbed on top of the muck heap, started to roar and throw manure around at the sight of the person he hated.

The boss attempted to drive the bull off the muck heap by approaching him holding a pitch fork with the prongs as far away from him as possible. When the bull showed that he had no intention of leaving his advantageous position and gave a further roar the boss turned and ran with the fork still pointing over his shoulder at the bull.

This was very apt since it was about this time when British cartoonists were making fun of the Italian soldiers' fighting capabilities by depicting them as 'advancing' by running away from the enemy with their weapons pointing over their shoulders, exactly like the boss running away from the animal. From a safe distance I was creased up with laughter.

This bull would never have hurt me or the cowman deliberately but I am sure he could have killed us by accident when he was let out into the yard to serve a cow.

The yard was parallel to the cowshed and reached by going through the barn. By the door from the barn to the yard there was a set of steps leading to the second storey. The procedure was to turn the cow out into the yard first but she was naturally confused since she knew she should have gone the other way and so I stood waving a stick to keep her in the yard. When I heard the cowman shout I knew that he had released the bull so I had to belt up the steps before the bull came charging from the cowshed. He would turn sharp right into the barn, slipping and sometimes falling in his haste. Anything in his way would have gone with him, including me.

In the end it was decided that he was dangerous and he was replaced by a roan shorthorn bull. I was uneasy with this one because when he was turned out with the cows and I drove them in or out I would find him following me. I had a feeling that if I didn't treat his ladies right he would turn nasty. I preferred him in front of me where I could see him.

The next permanent cowman was a younger man, probably in his forties and although he was small he was very strong because he had worked as a docker. I was really rather impressed with him. He had quite a lot of children and in my cheeky manner I mentioned it to him.

'Ah, well,' he replied, 'I hadn't been married very long when my wife took a fancy to somebody else in the village and moved in with him. I decided, right, I'm not having that. One wet night I went to this man's house and when he came to the door I pushed past him and went upstairs. My wife was in bed and I pulled her down the stairs and left. I didn't stop for her to get any shoes or clothes or anything. After that I thought I would give her plenty to do which is why we have so many kids. I haven't had any trouble with her since.'

He was a hard-working man and had plenty of stories worth listening to but he didn't quite have the enthusiasm that old Frank had had for the job.

The most embarrassing event in my life was due to the cows on this farm. No effort was made to back-fence the land so that the grass would have a chance to regrow after being grazed. The cows were turned out into the field closest to the buildings and then worked their way through the succeeding fields which ran alongside the lane. It was a very hot, dry summer and although the cows had enough to eat, the grass was parched and tough, not exactly appetising.

For some reason I had been to the far end of the lane, perhaps to take a horse to the men working on the arable fields, because I was coming back on foot. The herd spotted me. 'There is our servant,' they said to each other, 'Why isn't she looking after us properly?' Then they charged over to the fence and followed me, hurling abuse in a loud, rude manner. You could imagine the chant: 'What do we want? Fresh green grass. When do we want it? Now, now, now.'

This would have been okay if it hadn't been lunch time. The NAAFI headquarters was housed at Ruxley Towers nearby but there was no canteen there so staff would cycle down this lane to another government department for their meals. It seemed to me that there were hundreds of them coming towards me, falling off their bicycles with mirth.

Had I been older I would have joined in the fun but I was a very shy seventeen-year-old, and, try as I might, it is difficult for one small girl to pretend that she is not being followed by a herd of great bellowing beasts.

A few months before I was eighteen, I joined the Land Army. Since I was already doing farm work I had to join before I reached eighteen otherwise they would not have me. The idea was to get people into the industry, not to hand out uniforms to people who were already there.

I was horrified when I was told that I would be staying on the same farm. In the Land Army you could refuse a posting but, once there, you could not leave.

Fortunately, the farmer's meanness came to my rescue. Not only were my wages due to go up when I was eighteen but the Land Army told me that I was entitled to one weekend off each month. This combination was enough for the farmer to decide that a younger person could do my job. This earned him a place on the Land Army's blacklist.

Chapter 4

In The Land Army

I JOINED the Land Army in 1943 and stayed until it was disbanded in 1950. Since I was already working on a farm and didn't need any training it was quite some time before I met other members of the WLA.

We were visited at intervals by the 'Are-you-happy-in-your-work?' ladies – more formally known as the Land Army authorities. They also checked with our employers that our work was satisfactory. Two of these elderly volunteers organised a club for local Land Army girls at one of their big houses and my main memory is cycling home during the blackout and, more often than not, ending up in a ditch.

The Land Army was drawn from all walks of life. At one extreme were the girls who had led very sheltered lives. I remember one girl who had never even had to do her own washing and we had to show her how to do it. At the other extreme were the tough girls from inner city life with an answer for everything. Somewhere in the middle was the girl who considered herself to have been brought up properly because her mother had told her that she could go into a pub but not to hang about outside and wait for the men to come out. I was amazed. In those days, women who regarded themselves as respectable would never have gone anywhere near a pub.

Any class distinction that exists now is nothing compared to how it was before the war. There were very few middle-class ladies who would do their own housework and the ones who did would take great care to hide the fact. People who earned their living from a trade were looked down on.

My mother was once invited to tea with the mother of a child with whom I was at school. When the ice-cream man cycled past, the little girl wanted to go out and buy one. Her mother looked out of the window and said, 'Not now dear.' She turned to my mother; 'I don't want her going out when that child is there. His grandfather was a butcher.' It seems to me that this sort of person's opinion was in inverse proportion to their usefulness.

I used to get my dinner at the British Restaurant near where I worked. At the front of the queue was an over-cosseted, over-corseted matron who was fussing and complaining about the food and taking a very long time to decide what she would have.

British Restaurants were set up by volunteers to

provide a good cheap meal for workers without sacrificing their rations so I thought she had no right to be there at all.

Behind the matron were three Land Girls who were in a hurry because they only had limited time for lunch. Finally, they were so exasperated that one of them said to her, 'Oh, get out of the way.' This self-important woman turned to her and said, 'How dare you talk to me like that.' The Land Girl looked her straight in the eye: 'We talk to all our old cows like that.'

They were not the only ones to learn to speak up for themselves. I got home one evening to hear the hairdresser's assistant telling my mother what they both considered to be a shocking tale. One of her clients had gone into the fishmongers and complained that they had no flat fish. This was at a time when fishermen were risking their lives in the face of enemy action, so the young woman behind the counter replied, 'If you want it flat, put it through the bloody mangle.' Apparently this customer had never been so insulted in her life.

When I started working for the Land Army I was sent to work on a sixty-acre farm in Surrey where I was to be in charge of calf-rearing. It was one of four owned by the farmer, the other three of which were dairy farms. The main enterprise on my farm was a large pig unit, but I never got involved in that. My job was to rear the calves which were brought to me from the dairy farms at a few days old.

The owner of this farm appeared to bit of a wide boy and there was no shortage of strange rumours about how he had made his obvious riches. The calf pens were divided with iron railings which had been cut down by Government order to be used for the war effort. They had been sunk spikes-down into concrete and worked very well. However, they had been coated with lead paint which caused several cases of lead poisoning when calves licked them.

Not only had he 'won' these railings, but the calves were largely fed on condemned tinned milk and broken biscuits. How this was all obtained, it was not wise to enquire but the farmer's wife could open tins faster than anyone I have ever seen. The calves were Friesians and Guernseys. The former were much stronger and thrived on this diet but the Guernseys were delicate so I spent many hours helping them learn to drink from a bucket. During the whole time while I was there I only lost one calf and that was a premature twin not much larger than a rabbit.

My helper was a young man called Hubert. He had been invalided out of the RAF and was what we would now call 'a sandwich short of a picnic'. When air force personnel heard that he had been the Padre's runner they knew exactly what his mental capacity was. However, he was a very pleasant chap and liked working with me because I gave him a bit of responsibility. The older woman who had been in charge before I took over had only allowed him to clean out the pens and do the washing down.

One day, over our mid-morning cup of cocoa, Hubert asked me if I knew why he spat blood. I passed it off by saying that he was probably anaemic but realised that he had been invalided out of the RAF because he had tuberculosis. Antibiotics were not yet available and he had probably just been told to get an outdoor job. He should never have been working with cattle, or with anybody indoors. I think his parents realised this because he left within a few months. He was dead in just over a year. Every time I hear some of the old war-time songs I think of him because we used to sing them as we worked.

On this farm I met another rather odd character, a young man who drove a lorry between farms. At lunch he would often come and join us in the calf shed with a racing paper and study form. I think he lost more than he won but his really eccentric behaviour concerned his teeth. Although young, he didn't have a single tooth in his head and to eat he would put in his false ones. The rest of the time he kept them on the top of his head under his cap!

It was 1944, the time when Hitler's flying bombs, which we called Doodlebugs, came flying over Surrey. They were aimed at London but where they came down really depended on whether they were on course and when they ran out of fuel. They flew very low and so long as you could hear them and see the flames coming out of them you were safe. When their engines cut out you hastily took cover.

My landlady's daughter, Stella, was working in an office in Kingston-upon-Thames when one of the first ones came over. The staff ran to the window to watch it. Then, to their horror, it cut out, the wind turned it and it headed straight back towards them. Fortunately their only injuries were caused by cracking their heads when they dived under the furniture.

Stella was waiting to join the Land Army and was sent to Limpsfield to study dairying. This was where many Doodlebugs came over and the girls were told that if they were milking they must wait for the man in charge to give the order to seek shelter. Sure enough, one day a Doodlebug cut out overhead. Their instructor had a stutter but finally managed, 'Ger, ger, ger, get out!' Stella said they were poised like sprinters waiting for the starting gun but obediently waited for him to give the order.

We had six cows whose milk was mixed with the tinned milk for the calves. One of them had her calf out in the yard and in its struggles to get to its feet it fell into a deep, but fortunately dry, ditch. Hubert and I were standing at the top wondering how we were going to reach the calf when a Doodlebug cut out overhead. The only problem then was how *we* were going to get out of the ditch.

When the pig unit got swine fever the owner decided to sell the farm. I had recently passed my proficiency test and been promised a wage rise. When I told the manager that I hadn't had it he sent me to see the boss who gave me a five-pound note. I had never seen one before because it was equivalent to two weeks' wages. The boss offered me a job on one of his other farms but Stella had been posted to a Jersey farm in the same area and had told me about a vacancy. I decided that would be a better move.

The farm was situated on the Hogs Back between Farnham and Guildford. It was owned by a millionaire as a 'hobby farm' which meant he used it to offset income tax. Food growing was encouraged by any means. How much this farmer knew about farming was unknown to us but it didn't matter since each department was run separately by its own expert with sections for pigs, poultry, arable and dairy. All were at the forefront of new developments. For example, breeding bronze turkeys and growing maize as fodder.

I remember the bull on this farm very well. He was

called Beautiful Dreamer. Beautiful he certainly was but not much of a dreamer. In fact, like most Jersey bulls, he was rather belligerent.

We used to groom him every day and in order to do this we shut him into his loose box with food in his manger to keep him quiet. There were neck yokes through which he had to put his head to reach the manger and we were supposed to close them around his neck with chains operated from outside the pen. However, it was impossible to use them without barking your knuckles against the brick wall so we didn't bother.

One day the farm manager came round with a visitor while I was brushing the bull who, of course, thought that his territory was being invaded by a stranger. Pushing me out of the way he advanced on the door which the manager promptly slammed and bolted shut, leaving me shut in with Dreamer. Fortunately, he wasn't interested in me so I was able to unbolt the other door and slip out.

It has been said that there wasn't much recognition of what the Land Army had done once the war was over but I am not so sure that I agree. During the war, correspondence courses were available; I did one in dairy farming. There were also proficiency tests for which highly trained personnel would visit the farm and give practical tests to candidates. There were opportunities for those who were prepared to take them.

This was a very forward-looking strategy since at the time there were no differentials to ensure that more highly skilled workers should earn more than their unskilled counterparts. For the keen young person this was very discouraging. The only increase in wages was for those who worked longer hours than the general farm worker; it would be several years before different grades were introduced by the Agricultural Wages Board.

During the period between the end of hostilities in 1946 and the disbanding of the Land Army in 1950, women who intended to make a career on the land were given the chance to take an intensive course for a full academic year in order to qualify them to take supervisory or farm secretary posts. I went to Thriplow in Cambridge with thirty-five other Land Girls; it didn't cost us a penny and was first class.

We were divided into three groups, one of which would have to turn out and do some work at 6.00 while the rest only had to surface for breakfast at 8.00. We had lectures from 9.00 in the morning until lunch-time. There were three resident lecturers for crops, poultry and animal husbandry and a county machinery officer came in to teach us about tractors and implements. We also learnt book-keeping. A lovely old gentleman known universally as Uncle David came to demonstrate and teach bee keeping.

In the afternoon we were divided into two groups, one of which went on a farm visit while the other did more practical work with a view to looking at it from a supervisory point of view. At about 4.30 we would come back for a cup of tea. One of our jobs in winter was Brussels sprout picking and if you didn't have a mate in the other group you wouldn't get a cup of tea because your hands would be so numb you wouldn't be able to operate the tap.

After our tea we would write up the work of the day until supper time which was 6.30. After that we were free apart from those detailed to do the washing up.

We had lectures five days a week and on Saturday mornings were detailed to do domestic work, the same as you would in a youth hostel. From Saturday lunch-time until Monday mornings we were free. In the summer, about sixteen of us would cycle to Cambridge and take out rowing boats on the River Cam. We rowed as far as we could towards King's Lynn and swam on the way.

At this time Thriplow farms had three dairy herds of Jersey cows which were milked in different ways. The first group was hand milked, the second was machine milked into buckets and was the most efficient, and the third used a prototype parlour system. This system meant freeing each cow from the shed one at a time, putting them into a stand where they had their udders washed, moving them to the milking stand and then returning them to the cow shed. This took at least as much, if not more labour than hand milking, especially when you consider the complicated cleaning of the milking units. Girls on duty were allocated to one of these three units or to a neighbouring poultry farm.

Thriplow also had a breeding stud of Arab horses and I remember being given the honour of grooming a most beautiful copper-coloured stallion. When I had finished, the head groom said, 'You did quite well but you've got cows written right across your forehead.' I often wonder how he knew.

Our first warden was a very pleasant lady, although she knew all the tricks we would get up to. If you wanted to go out in the evening you had to have permission, otherwise you had to sign in by 8.30. One day she commented that all the names were in the book – but many were in the same handwriting.

After she left she was replaced by a real dragon. For whatever reason, this lady decided that we were getting a bit too friendly with the lecturers who sat at a different table and so we returned from one of our vacations to find that all the china which the lecturers used had been painted with multi-coloured spots in order to distinguish them from us. In addition to those detailed to do the washing up, a couple more girls volunteered to remove the spots with pen knives. It took three days but they never reappeared.

Although we worked hard we also had a lot of fun and often bent the rules and played jokes on each other.

Although we were not supposed to be out after 8.30 without permission, four of us had the top bedroom which was very useful because the fire escape came up to the window.

One of our four was a very staid person. We would say to her, 'Now Zoe, go down and sign the book for us at 8.30 and don't forget to leave the window open so we can get in.' She became a probation officer in the end; I suppose she felt she had experience after dealing with us.

We were not up to very much mischief; all we did was go over to the Nissen huts where girls over twenty-five had individual rooms and we spent the evenings there. Although we were well fed we were at that age when you could never be filled up and were always hungry. One of the girls had brought back an electric toaster and so we would go over, particularly on Sunday mornings, and have toast.

One day we saw the Dragon coming round on a tour of inspection so one of the girls shoved the toaster into an empty suitcase under the bed. We sat there waiting to the place to go up in smoke.

Another thing which was strictly forbidden was to bring alcohol onto the premises, so we liked to smuggle in beer. One evening one of the lecturers was teasing us about this so someone dared me to take some of the sharp serrated bottle tops and put them into his bed. I did.

The next morning he told us about this and said he was glad he didn't catch whoever did it. He had no idea who it was.

Next to us was another room with four girls in it and the bathroom was in-between us. On the next floor one of the bigger rooms accommodated eight girls who were our rivals even though in the summer we went out on the river together.

One night we came back to find that they had decorated our room with toilet paper and really caused chaos so we set out to get our own back. Our beds were camp beds and on the same floor was a sick bay with a key in the lock. One night when they were out we took down their beds and all their bedding and locked it in the sick bay.

Lights out was at 10.30 and very strictly observed. If you were not quiet by that time the Dragon came round and there would be trouble. We sat giggling on the stairs while our rivals crept about trying to find their bedding. In the end we took pity on them and told them where it was.

The only person who didn't appear to have fun was a poor, thin, meek little creature called Muriel who shared our bathroom.

Normally we would all pile in there or, if the door was locked, we would hammer on it and say, 'Come on out, you've had long enough.' If it was Muriel in there she would come out and say very apologetically in a whisper that she was sorry she had kept us waiting. In the end people would say, 'Don't knock on the door, it's Muriel in there.'

We felt sorry for this girl because she was always by herself and working at her books so one Saturday we asked her to come out with us and enjoy herself. She replied rather sourly that we were not there to enjoy ourselves.

My father always used to say, 'There was once an old woman who wasn't happy unless she was miserable.' It seemed that Muriel was the same, determined to always be a downtrodden failure. At the end of our course we had exams and tests; she was probably the only one who didn't pass.

Muriel reminded me of a boy who worked on a farm near Bubblewell. He had a reputation for being a hard worker and so for a while I entertained the idea that we could go into business together and get a small bit of land. I came across him once when the fair was in town and thought it would be a good chance to get to know him. 'I'm going to the fair to roll my pennies and see if I can win anything. You coming?' I said. He snorted, 'I haven't got time or money to waste.' It wasn't so much what he said but the gin trap way his mouth snapped shut that made me immediately change my mind about working with him. He and Muriel were just alike, they even looked similar: pale, thin and hunched up like dead flies in a spider's web.

Chapter 5

The Land of Tre, Pol and Pen

WHEN our course in Cambridgeshire came to an end, job adverts for farms nationwide were posted on a board and we were encouraged to apply for them. I spotted one in Cornwall. I had been to places like Brighton and Yarmouth before the war but never anywhere as romantic as I thought Cornwall was, so I applied for the job and got it.

It turned out that I had made no mistake in choosing Cornwall which proved to be a fascinating county although the way of life was different from what I had been used to.

When I arrived at Polgrain Farm on the Caerhays estate in the autumn of 1947 I immediately marked myself out as being 'one of they odd 'uns from up country'. There was a gap in the trees at the bottom of a hill and I asked, 'Is that the sea down there?' Of course, the last thing that Cornish people wanted to do was get near the sea: it was regarded as very dangerous and they only went near it if they had to.

None of the women and only a few of the men could swim, even though there was a lovely beach there. One ex-fisherman told me that they didn't learn to swim because if they fell in it would only delay their death and they thought it better to die quickly.

I discovered very soon that Cornish people at that time were very parochial. At one point I was talking to a man about how the world's population was increasing. 'That it's not,' he disagreed. 'There are fewer people in Portloe than when I was a boy.'

Soon after I arrived it was harvest time and I was catching sheaves on the rick. Normally on that farm there were the farmer, one other man and me. During harvest anyone who had some free time would come and help and one of these helpers remarked, 'I hear they have a new man at the Barton.' In Cornwall the largest farm in the village is often called the Barton. 'Oh yes,' his friend replied. 'He has a very hard-working wife, a north country woman.' I was just about to ask if she came from Yorkshire since my father was born there when the first man replied, 'Yes, she comes from Barnstaple.'

The first time the farmer went to market he wanted to give me something to do while he was away so he sent me to get the horse shod at the nearest blacksmith's, two or three miles away. Some retired old men who were gathered outside the smithy moved back with some very off-putting looks to let me through.

I stood at the horse's head and through the pungent smoke as the farrier tried the hot shoe on the hoof and the hissing steam as he plunged it into the water to cool, I heard one of the men say, 'You get some rum-uns from London.' It seemed like, to them, there was just Cornwall and that everything on the other side of the Saltash Bridge was London.

I soon discovered how hard it was to remember the place names. In the rest of England if you just remembered the first syllable the rest seemed to follow quite easily but in Cornwall all the places seemed to begin with 'Tre': Tregony, Trevaras, Trevose, Trevarrick, etc.

Even worse than remembering the place names was understanding some of the phrases. The Cornish language had not been spoken by ordinary people since the seventeenth century, but some phrases such as 'put the slips on the arishes' and 'pare the mowhay' had managed to survive. I soon learned that 'arishes' meant stubble and that 'slips' were the young pigs who were turned out onto it to pick up the fallen grain. They also spoke of 'buss' calves meaning calves which had been suckled on the cow and it was 'make out the light' instead of 'put out the light'. They didn't ask 'Where do you come from?' but 'Where do you belong?' which I was thought was much nicer.

Soon after I arrived we were building a rick in the yard. As the boss went off to fetch another load he said something that I didn't understand. Since the field was some distance from the yard I had to wait between loads and so as not to waste time I got a sickle and started to cut down the nettles around the yard. When I told the boss what I had done I expected recognition for my initiative but received none. It turned out that he had said that I was to 'pare the mowhay' and I had inadvertently done as he asked. 'Paring' meant trimming or cutting with a hook and the 'mowhay' was the stackyard.

In the end I could pass myself off as a native which meant that when I went to other parts of the county I was treated differently from 'they visitors'.

It was while we were working on this rick that I found out why Tom, one of the general workers, wouldn't speak to me. The sheaves had a lot of thistles in them and he wasn't wearing gloves. I thought he was very stoical not to say a word when he caught hold of some. In the end he said to me, 'Is it true your father is a vicar?' I said, 'No, whatever gave you that idea?' and he gave me a big smile. Then he spotted the boss approaching and said, 'You bugger, why did you tell me she was a bloody vicar's daughter?' The boss creased up with laughter. He had told Tom, who couldn't open his mouth without a stream of expletives escaping, that he was on no account to swear in front of me. Once he found out it was OK he was quite chatty.

The only other swearing I can remember hearing in Cornwall (they were mostly strict chapel folk) was when a vet visited the same farm with a small son in tow. While his father was examining a cow the little lad asked, 'Why won't the bloody cow stand still, Daddy?' The vet stopped looking at the cow, 'What did you say?' From the tone of his father's voice the child knew he'd said something wrong but had no idea what so repeated the question hesitantly but word for word. His father chased him around the yard which I thought was very unfair. How was he supposed to know?

Polgrain farm was a 150 acre mixed farm and not a good one to work on. The farmer was a smart little man with a lot of good points but a lot of bad ones as well.

He used to do things like buying in potatoes that were intended for stock feed. They were deemed unfit for human consumption and dyed purple but he bagged them up and sold them to unsuspecting old ladies.

At the time we had tests for tuberculosis in cows and if one tested positive you had to get rid of it. He would get me to take any cow which he thought would fail to an out-of-the-way field until the tests had been done. If the animal was a reactor it could spread the disease to the whole of his herd so although he thought he was being clever he was really cutting a rod for his own back.

On the other hand he was a very kind man who always took the trouble to open the landing window for a butterfly which was trying to escape into the sunshine. The cleaning lady used to squash the poor things.

I lived in for the first few months until the boss's wife was expecting her second child. Both of them were roly-poly people and she, especially, reminded me of a Dickens character. He had the dark hair and blue eyes of a true Celt, his ancestors probably having lived in these parts for hundreds of years.

Good eating was no doubt the cause of their plumpness. One of their favourites was real clotted cream spread on a hunk of bread and topped with golden syrup. We had many a discussion about whether to put the cream or syrup on first! The missus would also make a lovely sponge cake which we would have for Sunday tea in the front parlour.

As chapel people they took their turn to entertain the visiting preacher. One of these was a very earnest young man. One Sunday, when this man was there, the boss and I were washing our hands in the kitchen sink before going through to tea. The boss said to me, 'You wait, I'll make you laugh.'

I was sitting there, trying to look angelic, when the young preacher remarked, 'So you've finished another day's hard work.' 'Yes,' said the boss, with a sigh and looking at him with innocent blue eyes, 'Another day's march nearer home.' It was unfortunate I had a mouthful of cake. Crumbs exploded all over the table.

Really, the boss hadn't wanted someone like me: he wanted a glorified yard boy who could milk the cows quickly. He wouldn't allow me to learn any new skills because while I was learning I would be slow. During the second harvest I was there we had a college boy on holiday who came in to help. He was too lazy to put the stooks up properly and they promptly fell over, so the boss put him on the binder and wouldn't give me a turn. He never let me drive the tractor either, keeping that for himself.

He also got annoyed when I asked him questions. I assumed that he would know why he used three hundredweight of fertiliser per acre on a certain field rather than two or four but it is likely that he was just doing as he had been told and had no idea what the effect of more or less would be. He did not appreciate being asked.

I really did have the knack of saying the wrong thing to this farmer. When I was reading about the history of the area I came across the story of a smuggler who had known the coast 'like the back of his hand.' He shared the unusual surname of Kerkin with my boss whom I told about my discovery, expecting him to be pleased that this famous man was almost certainly an ancestor. He was not.

After I left, the farmer's small son gained fame for having killed a bull when he was just four years old. The bull was loose in the yard when the farmer's son found him there. Remembering that bulls like eating crushed grain the little boy opened the door to the feed mixing shed and called the bull in. The bull must have thought he was in heaven which was ironic because by the time he was discovered he had eaten so much it killed him.

It is strange how animals which can be regarded as dangerous can be so gentle with children. When I think of that little boy and the huge bull I am reminded of a story which a lady told me about her toddler grandson and Tamworth pigs. I commented that they were considered to be nastier than other breeds but she told me that she thought it was how they were treated.

She and her daughter were in the kitchen preparing a meal when her son-in-law came in. They asked him where the little boy, Andrew, was. He said, 'He's not with me. I thought he was with you.' Panic! They searched the house and all the buildings, shouting and calling, but to no avail. The boar lived in a small paddock just across the yard from the back door. Normally, if anybody went outside and called, the boar would come trotting over for a titbit but on this occasion he was lying outside his shed at the far end of the paddock. He just raised his head but made no effort to get up. His owner thought perhaps he was ill and couldn't get up, so she went over to see. There was little Andrew lying fast asleep with his head on the boar's shoulder. The boar remained still until Andrew was lifted up. Thereafter, there was a very strong bond between the little boy and the boar.

The good thing about working on Polgrain farm was that once I moved to Cornwall I began to do more general field work such as hoeing, paring and muck spreading instead of spending all my time in cow sheds. It may seem that you do not need any skills for this type of work but you would be surprised.

Take hoeing for instance. The difference between someone who was skilled and someone who wasn't was not only the speed of the work but how accurately they could thin crops such as mangolds which were grown in the West Country. We used to hoe them twice while they were growing and the boss always said that it was not always the one who does it quickest the first time who does it quickest the second time because the weeds that were missed have grown a lot bigger.

The essential thing when using a sickle or scythe was to get the blade really sharp and take a full cut each time while being careful not to dig the point into the ground. I never got particularly good with the scythe because we only used it in places where the grass cutter couldn't get to, or to get a bit of green fodder for the bull.

In Cornwall they were very keen on doing jobs using both hands alternately where possible. Of course, hooks and scythes were designed to work only one way but you could use a hoe with whichever hand you preferred. At first it felt very awkward to change hands but it was worth persevering because it saved a considerable amount of backache. You could always tell when I was getting tired because I changed hands all the time.

On this farm I worked with an interesting mare who would play me up at any chance she got.

I often took her carting a mile or two from the farm through the narrow Cornish lanes. The first time I did this the boss warned me that she would probably stop and refuse to go any further when she felt that she had gone far enough. When this happened I was to cut a little stick from the hedge to encourage her. Apparently, waving it where she could see it would be enough.

Sure enough, after a bit she stopped and without thinking I jumped down to cut a stick from a young ash just ahead. When I heard a noise behind me I was horrified to see that 'Madam' had managed to turn round. However, because the lane was so narrow, one of the wheels was right up on the bank and the cart was in imminent danger of turning over. This could have been disastrous. The cart could have been smashed and the horse badly injured. Fortunately, the mare was so strong that momentum kept the cart upright until it was on the road again, and then she set off for home.

She was going much too fast for me to catch her so I vaulted over the nearest gate, set off across the fields and was waiting for her when she came around a corner. Her eyes nearly popped out of her head and she put the brakes on so fast that the cart nearly knocked her feet from under her.

I made sure I never had any more trouble with her and it was a lesson in thinking about what you're doing. I should never have gone ahead of the horse when home was behind. The ambition of nearly all horses seems to be to get home as soon as possible.

Another thing that this mare did, I wouldn't have believed if I hadn't been driving her at the time. I had allowed the rope lines (reins) to become slack while we were crossing a field because the ground was rough.

As with most working horses at the time, she had a cropped tail. She was holding it up and waving it from side to side. I assumed that she had some sort of irritation but actually she was feeling for one of the lines and as soon as she found it she clamped it under her tail. I had never realised how prehensile a horse's tail is. I could not get the line back and she was able to set off for home. In the end the only way I could stop her was to jump onto her back and grab the lines by her head.

Another day she decided that she had done enough

work and so sat down on the hay rake refusing to budge. I found using the hay rake difficult enough without her being awkward. It consisted of large curved metal tines, designed to collect the hay from swatches into rows called windrows. Then it was either swept by tractor to a stationary baler or baled directly. In order to drop the hay in the windrow you had to operate a pedal and lever together. Since the rakes were designed for men to operate and I was only five foot three, no matter how hard I tried, the windrows would wiggle about making it difficult to sweep or bale the hay afterwards.

I vividly remember my first experience with a strange remedy when I found a yearling heifer breathing heavily. I suppose the boss was away because I phoned the vet who suspected pneumonia. He wasn't able to come for an hour or two so told me to apply a mustard plaster to the animal. Of course, I laughed but he assured me that he was serious.

With a good bit of giggling, the farmer's wife and I made a concoction of mustard, flower and water which we spread on brown paper and stuck to the heifer's chest. With some sacking and a needle I parcelled her up and we hoped for the best. I was still convinced that the vet was pulling my leg and when he arrived would fall about laughing and get several free dinners out of the story. He turned out to be an elderly man whom I hadn't met before and who assured me that the mustard plaster may well have saved the cow's life!

Polgrain farm rented some grazing land a couple of miles away and the boss would send me over there with three or four cows. My helper was a sandy haired bitch and she was brilliant.

I found that it wasn't difficult to drive a herd of cattle if they knew where they were going but that a few cows with no idea where they were going would make great efforts to rejoin the herd. We would set off up the hill to the crossroads. The cows then had a choice of four directions and would usually take the wrong one so I sent Sandy to bring them back. We repeated the exercise until they went the right way. We did the same at every crossroads, of which there were several on the way.

The boss had a habit of losing his temper and if someone shouted at me I would shout back, so we had plenty of rows. We got over all but the last one.

It wasn't long after we had finished harvest and there was another girl on the farm by this time. She and I had been picking up the potatoes which had been left behind by the potato spinner. The boss had the harrows behind the tractor and was breaking down some other land but kept popping back over the potato ground to see if any more would come out. He didn't give us time to finish before he was back again with his 'hurry up, get on with it' attitude.

After harvest I was not at my best and really needed a rest so in the end I lost my temper. I had a bucket in my hand and am afraid to say I threw it at him. That was the end of that relationship.

Chapter 6

St Mawes

AFTER my dramatic exit from Polgrain Farm the Land Army sent me to a farm near St Mawes to do three months' relief work so that the farm family which consisted of the farmer, his wife and their daughter could have holidays.

It was a sixty acre farm which was a reasonable size in those days and the only person who did any work was the wife. The daughter was allowed to live there free of charge on the condition that she helped with the milking which was unfortunate for me because she was also machine mad. If she heard the tractor starting up she would come running out of the house to drive it, so I never had a chance to do any driving there either.

The farmer was the last man who should have been farming. He had inherited it from his parents but didn't like the job at all.

When he went on holiday for a month he told me that if I finished the work with the cattle I could dig up the half-acre of carrots and make a clamp. Well, they were rubbish, tiny little things and full of holes – carrot fly I presume – so certainly no good to sell to the public. When he returned and I asked him why he grew them, he replied, 'Oh, I don't know, I always do.'

In addition to his lack of forethought, he was very lazy when it came to doing anything outside and, as they always say, the lazy man makes the most pains.

Several tons of grain would be placed over the hole in the barn floor which led to the grinder. The idea was that it saved having to fill it as normal and this made perfect sense when the grinder was running but if it developed a blockage, which it did often, there were tons of grain to move before we could fix it.

The same problem was true of his attitude to the cattle. He said, 'You don't want to have to move the straw from the rick yard for the young stock, let them in to help themselves.' Well, of course, they wasted a tremendous amount, pulling it out and trampling on it. After a while the rick came down because they were eating at the same height all the time.

In contrast to his lack of enthusiasm outside, he was a very artistic man. One day he took it into his head to have a go at icing a cake. It was a wedding cake, although his daughter was not getting married. I often wonder if he made a cake for her when she did because he made a beautiful job of it with twirls and curls and loops.

The only person that we really missed when she had her holiday was the farmer's wife. She was the sort of woman who had her washing done by 8.30 in the morning, totally unnecessarily really, and it meant that she let other people get away without pulling their weight because she did so much. She was also a fantastic caterer. Early in the morning she would bring us each a thick slice of home-made bread and butter and a mug of tea. When we came in after milking we would have breakfast which consisted of porridge cooked with creamy milk, a fry-up and more home-made bread and butter with marmalade. I couldn't eat like that now if I tried, and it certainly wouldn't do me any good!

On the farm there were three sources of energy (apart from the farmer's wife): a horse, an ancient car and an equally geriatric tractor. Every week the boss would go to market in Truro on the King Harry ferry because it meant that he got off the farm for the day. We would hitch the horse to the car and pull it along the farm track to the tarmac road. There was a convenient slope down to the ferry so we would get behind the car and push him off. Hopefully, after a few coughs and hiccups it would start and off he would go. However, if it wouldn't start before he got to the bottom of the hill we would have to plod down, hitch up the horse, and start all over again.

The tractor was too heavy for the horse to pull so if

we wanted to use it we had to start the car and hitch that to the tractor!

The horse in this exercise was a vigorous but well-mannered gelding known as Duke. One day he decided to take himself on a tour of St Mawes.

I had not been told that on certain days the Navy did gunnery practice in the bay. Once you have backed a horse in between shafts you have to let go of his head in order to lift the shafts and it was at exactly this moment that the Navy chose to let off the first salvo. Duke set off at a gallop up the farm track towards the road. Once he calmed down he pricked up his ears and trotted along looking from side to side with the harness bouncing on his back. He was obviously enjoying himself. Fortunately it was winter and there was little traffic on the roads because I found it so funny that I had difficulty catching him until he finally decided to stop.

It was a lonely job because there was only one other employee on the farm, a part-time chap. He loved driving the horse and would race him about with great loads of kale. He was not a very bright fellow. There were three small cow sheds with a feed bin in each which it was his job to fill up. I would have to be very careful not to leave the scoops in them otherwise he would bury them.

On my days off I would go across to Falmouth on the little boat. Some of the time that I was there it was very rough weather, but the boat would go so long as they could tie it to the quay. At first I made the mistake of going downstairs with the local people. Although it was only a twenty-minute trip the currents met in the middle and the waves were enormous so I felt very ill by the time I got to Falmouth. I discovered very quickly that if I stayed upstairs and hung onto something I got on quite well. In fact, I rather enjoyed it.

Once my three months were up they asked me if I would stay on. However, the Land Army regulations meant that if I accepted I would not have been able to leave later. I decided it would be better if I were to leave straight away.

While I was waiting for the Land Army to give me another posting I had a phone call from the uncle-in-law of the farmer at whom I had thrown the bucket. At first I could not understand what he was saying because he had a strong accent and a lisp. I eventually gathered that he was asking me if I could type. Fortunately, I had been to typing and book-keeping classes when I was working on the calf rearing farm in Surrey. He said that if the Land Army agreed he would give me a month's trial. This would prove to be the most interesting job I had had to date.

Chapter 7

The Barton

THE man who phoned me up was the manager of Home Farm on the Caerhays Estate near St Austell in Cornwall. As the biggest farm it was known as the Barton as was usual in Cornish villages. The manager, who went by the interesting name of George Thomas Kneebone, also had responsibility for the whole estate which stretched for miles in either direction and was a fabulous character although very much disliked by many people. The foreman's brothers told me years later that they regarded him as only one degree down from Hitler. The owners of the estate had another in Cornwall as well as one in Scotland whence came the Highland cattle which led our quiet South Devons astray. Peaceful afternoons in the office would be interrupted by irate neighbours informing us that our cattle were trampling their crops.

This farm was about 400 acres, which was a big farm in those days and certainly bigger than I had come across before. There were a cowman and a girl who worked with him, a stockman, a shepherd, two tractor drivers, two horsemen, a foreman, two retired men who worked a few hours as needed, and me.

As well as working outside I did the book-keeping and any typing that was required. The typing was not very onerous since we wrote very few letters and just a few invoices. The book-keeping was also only a part-time job since wages were paid fortnightly. There were forty-two people on the books including all the castle servants, gardeners and estate workers such as the blacksmith, stonemason, painter and carpenter.

I began my day at 7.00 and helped with the milking until 10.00 when I would go and see if there was any paperwork to do. If there was, I had until afternoon milking time or until I finished. If I didn't have much to do inside I would ask the foreman to give me something to do outdoors.

From a husbandry point of view the farm was extremely well run and everything had to be done exactly right. In order to keep down the weeds, even the corners of the fields were ploughed out by a horse if the tractor couldn't get in. However, this made for problems when harvesting with a machine because someone had to go in with a scythe. Everything had a right and wrong way to be done, even down to how you placed the brush after scrubbing the dairy. When we had finished with it, all the machinery was cleaned and greased or oiled. Even now, if I see machinery which is just left to go rusty I don't feel good about it at all.

At the Barton I met some very interesting characters, one of whom was our stockman. He had been crippled as a boy when he had been leading some horses in a binder. As he was turning the horses one of them stood on his foot and dislocated his hip. Although he had had very little treatment I don't think he suffered any pain but he moved forward with a snapping movement. His head went forward and then would snap back up to propel him along. He was very thin and wore an enormous belt. I always had a suspicion that when he took it off at night he was in two halves!

There were two horsemen who were entirely opposite in character. One was considered competent to take responsibility if the foreman was away but he hated it. The foreman only had one week away at a time but this man would go around wringing his hands saying, 'I wish Jim was back.'

The other horseman, Steve, was completely laid back and took no notice of anybody or anything. Sometimes Mr Kneebone would say, 'Come on Steve, you've got to do a bit better than that.' Steve would stutter, 'Oh, all, all right.'

Mr Kneebone would continue, 'It's not all right, Steve.' But Steve would just repeat himself: 'Oh all, all, all right.' It was all you could ever get out of him.

The story went that several years earlier they had been expecting the Germans to invade at any time. When the highly strung horseman saw what was probably anti-aircraft fire and mistook it for parachutists, he ran around shouting, 'The Germans are coming, The Germans are coming.' I don't know if they got around to ringing the church bells to warn people.

When this happened Steve was out with the thistle cutter which was only used by those who were sufficiently intrepid since it was kept for the slopes which were too steep for the tractor. It was really just a few bits of wood from which a chain drove something a bit like a chaff cutter set horizontally. We called it the *Daily Mail* since it came out every day in summer. Steve went up and down the precipitous slopes, shouting at his horse who, equally, took no notice of him.

Steve had stopped for what we used to call 'crib' – his mid morning break – and was sitting under a wall when the gamekeeper's wife came running past shouting, 'Steve, Steve, you'd better go home, the Germans are coming.' Steve stayed where he was: 'I, I, I'm having my break.'

If the Germans had come into the field he'd probably have carried on and taken no notice, and they would have taken no notice of him. Everyone else would have been tripping over themselves and probably got shot. Steve would have gone on quite happily.

It was on this bigger farm that I came to realise that workers came in three distinct categories. First there were those on which the industry depended: stockmen, shepherds, head cowmen and foremen. This type of worker was truly independent; most of them controlled their own lives to a great extent and loved their work. A shepherd would be asked how many acres of roots he wanted and about the grazing then he would go out in the morning and do his job. No-one interfered with him. The responsibility of a stockman depended on the individual farm and the generosity of the employer but a real stockman worked out breeding and feeding plans for himself.

In those days there was plenty of room on farms for the people in the second group, those with a pretty low mental capacity, because there were many more capable people to keep an eye on them and lots of fairly simple repetitive work such as hoeing, milking and cleaning stables.

The third group was those who could have taken more responsibility but didn't, even though many of them were highly skilled, interested in their jobs and worked hard. Whether they thought it was not their place to put ideas forward, were afraid of getting the blame if things went wrong, or just could not be bothered, I was never sure.

When I went to work at the Barton and was sent out

to work in the field there would be men with years of experience who would ask me how to tackle a certain job. Of course, I was inexperienced and would get it wrong. When Mr Kneebone asked why we were doing it like that they would say, 'That's the way she said to do it.' I learned a lot this way because Mr Kneebone always had to teach me the right method so that the others would follow. However, I never understood the attitude of these men.

One such man was an old widower called Alec. He was a rather unsavoury looking character. How often he washed his shirts and underclothes I don't know but his outer garments never even had the mud scraped off them. There were thick layers of farm dirt and spare bits of meals cemented together with dribble because for some reason his conversations seemed to need considerable lubrication.

He had a surprisingly good sense of humour. Sometimes he would say, 'You can come to tea anytime you like but bring your own food and then you'll have what you fancy.' Or if he knew I was going home he would say, 'Bring me back a woman from London.'

The poor old boy must have been very lonely. He put saucers of milk out for the big black slugs that we got around there and said they were his pets. He also used to keep pennies so anyone needing change for the nearby phone box would go to him.

Alec had been captured in the First World War because he had stayed with his mate who had been wounded. After the war he had gone to Canada and his main recollection seemed to be a landlady who had fed him banana sandwiches. He said, 'How can a man do a day's work on banana sandwiches?'

He could certainly not drive a tractor, would not drive a horse and very sensibly didn't learn how to milk cows – so he just did odd jobs. He liked turning handles such as the one on the potato riddle which we used for our small amount of potatoes.

In the summer he would say to the foreman, 'Are the mangolds ready to hoe yet?' In the West Country we grew mangolds, swedes and potatoes but no sugar beet and each of us would be given a certain number of rows to hoe. Alec was given more because he spent most of his time hoeing and paring. The stone walls in Cornwall were filled with earth and grew a considerable amount of weeds which had to be cut back before harvest. This was Alec's main job.

In the winter he would also spend a lot of time spreading manure from heaps in the areas where the muck-spreader could not go, and when it came to lunchtime he would just sit down under the nearest heap. Once he asked the foreman to buy him some tripe. Afterwards the foreman said, 'You should have seen old Alec today. He didn't know that tripe needed cooking, he just put it between slices of bread and was trying to eat it. It stretched like rubber until it was two foot long, but he still didn't make any impression on it.'

The complete opposite of Alec was a tractor driver called Matthew who thought himself superior to farm workers and called himself a fitter. 'Oh yes,' snorted the foreman. 'Calls himself a fitter but doesn't even know which way to turn a nut.' Now, to some extent that was unfair because we ran Case tractors which were an American design. Very cunningly, on one side the wheel nuts tightened the normal way but on the other side they tightened the opposite way so that when the tractor moved forward they would not work loose.

In spite of Matthew's immense self-confidence he was not the brightest fellow and had no skills. He was in his thirties and would fly around saying how he had to carry all these old men. What he didn't realise was that they could do four times the work that he did without appearing to make any effort.

One day I had inside work to do until after lunch so the foreman told me to come out after that and tell Matthew that I would replace him. It was springtime and we were bagging up corn, partly for sale and partly to be sown on our own land.

We had an old-fashioned winnowing machine which was placed in a large barn where plenty of wind blew through. Since it was not mechanised Alec turned the handle. The undressed grain went into the hopper on the top then dropped through various screens to sift out small grains, dirt and weed seeds. The clean grain had to be accurately weighed on the machine installed at the other end of the barn up some steps (the most convenient place for it).

Matthew had been placing a sack under the winnowing machine, filling it to about the right weight then taking it to the weighing machine on a sack barrow where he had a bucket of spare corn to correct it. Then he tied the sack and brought it back to stack by the winnowing machine before beginning the whole process again. He was stripped to the waist and sweating profusely. When I arrived he asked, 'How are you going to manage?' However, we had our orders so he left to prepare some land for sowing.

I said, 'Come on Alec, help me move the weighing machine.' It was a typical potato weigher with handles tucked under the platform and wheels underneath so we moved it over to the winnowing machine. Once we had put the sacks on the hooks, all I had to do was wait until the weight was right then move the sack into the bin. I said to Alec, 'Why didn't you tell him?' He replied with a big grin, 'I wanted to see him run.'

After that I was sure that Alec had known that what Matthew was doing was stupid but would not have dreamt of telling him. Whether he just didn't think about it or really did want to see Matthew run I don't know.

Matthew gave me the biggest fright of my life. He broke his wrist working on the seashore and so was absent from work for a while. He lived in the village about three miles walk away and as I thought that he would probably be wanting his wages I told Mr Kneebone that I would take them along in the evening. I set out in daylight but it was the time of year when it got dark at about 7.00, so after I had stopped for a cup of tea with Matthew's wife I left for home in the dark.

On the route home there was a steep rise on the hill between Portluney and Caerhays which in a car we were lucky to manage in second gear. It twisted and turned with a tunnel of trees over the top.

Suddenly I heard the sound of hob-nail boots behind me. What was really frightening was that the faster I went, the faster they went. I was pretty fit in those days so I walked as quickly as I could to get away from them and finally arrived at home gasping and with my heart practically bursting out of my ribs. I shut the gate firmly and waited to see who it was.

It turned out to be Matthew who said, 'Oh you do walk fast.' I could have murdered him! All he had had to do was call out and I would have recognised him. You can imagine how I felt although I never said anything to him.

Apparently Matthew's wife had said, 'You wanted to go and see Mr Kneebone and in any case it's not very nice for Sonn to walk all that way by herself. You go and escort her home.' (I was known as Sonn or Sonni in those days as Sonia sounded much too grand!)

In a situation like that your mind runs riot and I couldn't help remembering the ghost story of Trevanion's dog.

Although I have since discovered that there are several versions of this story, I always imagined a great, grey shaggy wolfhound or deerhound waiting for his master who had deserted him when he had run away from the bailiffs.

I was not the only one with this irrational fear. A gamekeeper warned me never to joke about it. He told me about a time when he and another man were waiting in the trees to catch poachers. He heard panting and rustling in the undergrowth and realised that his dog, which he had left shut up, must have got out and come to find him. Foolishly he nudged his companion and whispered 'Trevanion's dog!'

The gamekeeper told me that his companion was so terrified he thought he was going to die.

I got my revenge on Matthew although it was quite unintentional. There was a very steep piece of land running up from Portholland that we put to use by growing conifers. The land was planted by contractors, but for the first couple of years the weeds had to be trimmed by hand to give each pencil-sized tree the chance to get going.

When we advertised for someone to do this job the only answer we got was from a student, and although I would not say that Mr Kneebone's remarks were unprintable since he was not given to swearing, they were very caustic. I agreed with him. This was not the job for a soft-handed, bookish youth unaccustomed to manual labour.

It was decided that the farm and estate workers, or anyone else could have a go in the evenings so long as they brought a sickle. We had to turn up at 6.00 and stay until 9.00 so that the gamekeeper just had to book down names.

I was always looking for a way to earn a few extra pennies so I presented myself on the first evening. The gamekeeper looked down his nose and quite grandly informed me that I was quite welcome to have a go, but of course he would not expect me to keep up with the others. Poor innocent man, he did not realise the fury he was unleashing and I set out to chase him up and down until I had squashed his male chauvinist attitude.

I had my own little hook that I kept razor sharp, especially the tip which was the most difficult bit. We set off with the gamekeeper taking the first row and the rest of us spread out at safe intervals. If the gamekeeper saw that you were kept waiting by the one in front of you he moved you up. After a few trips I had worked my way up to the front which was a big advantage because it meant you could have a rest while the others caught up and, more importantly, had time to sharpen your hook. As soon as the last man arrived we would turn around and march down to start again.

Naturally, Matthew was determined to prove that he could do more than anyone else and so kept up with the front. But this was not work that he was used to and he had no stamina. We always said that tractor drivers were lazy. After a few evenings he stopped turning up – not only in the evenings but in the day as well. He had had to take to his bed.

Chapter 8

Beasts on the Barton

I MET several real characters among the people I worked with at the Barton and there were a few among the animals as well.

The Barton was the farm where I had the most to do with horses because there were three. The first was a very obliging little grey mare called Molly who would stand exactly where she was left and come when called. We had a lot of steep land, some of it ending in a cliff edge. A difficult or lazy horse may have let whatever it was towing go over the edge but Molly was completely reliable so we always used her in these situations.

The second horse was a large grey gelding. When you were leading him and wanted him to turn right you would push his head away from you and he would move his whole body to the right but when you wanted to turn left it was a different matter. When you pulled his head towards you he would obligingly turn it in your direction but the rest of his body would carry straight on. This caused me quite a lot of problems, banging into gate posts and chipping corners off walls until I got the hang of how to get the idiot to do the manoeuvre properly. One day I mentioned this to one of the men who remarked, 'Of course, he's a man's horse.' 'Yes he is,' I replied. 'He's just as stupid as a man.' They got used to me in time.

The third horse was called Lion and can only be described as bright red. Wherever you took him you had to take a 56-pound weight as well. Most horses will stand still while you load and unload but Lion had different ideas. His homing instinct was more developed than most so as soon as you let go of his head he'd be gone. We would put the weight on the ground and attach it to his bridle with a piece of rope. That put an end to his game.

If I was taking a horse home without an implement I would try to get up and ride to save myself the walk. One day I made the mistake of trying to ride Lion. As soon as I was astride he set off like a rocket, with me hanging on and hoping for the best. My main fear was that he would slip since his metal shoes were not made for galloping about on metal roads. As we got nearer the farm it dawned on me that the big gates would be open and he would charge through. This would be fine if he took it in the centre but if he cut the corner my leg would smash into the granite gate post which would most likely put paid to my farming career. I considered jumping off before we reached the gate but at that speed a landing on concrete would have meant a serious injury, so I clung on and hoped. We got safely through but that was the first and last time that I ever tried to ride him.

Horses were not the only working animal I dealt with at the Barton. There were also a couple of memorable dogs.

The best dog I worked with was an aptly named collie called Nip. He belonged to the foreman, Jim, and had a reputation for not letting anybody call at the house. When Jim's brothers called they would peep over the wall to make sure Nip was not around.

When I first arrived at the Barton to lodge with Jim's family I came on one of the two buses which ran each week and it arrived a little early. Without knowing Nip's reputation I marched straight up the path to find him stretched out over it. I put my suitcase down and stopped to talk to him. When he gave no sign of moving I stepped over him.

When Jim's wife came to the door she said, 'Oh, we didn't hear the bus. Where's the dog?' I said he was on the path and she was amazed because he had let a stranger pass. I am sure it was because I didn't know how difficult he could be and so was not nervous. After that Nip and I became firm friends and he would accompany me, not only when Jim told him to for work, but on evening walks as well.

Nip was well aware of the routine when there were visitors. We always knew when the baker was coming because Nip would hear him first and go into the wash house, ready for us to shut the door on him.

Next to the wash house was an outside lavatory where one night Jim left a newly acquired cockerel in his crate. Somehow the bird escaped and fell into the pan where, waterlogged and exhausted, he remained until morning. When Jim's wife's older sister, known as Auntie Fan, went out there in the morning we heard a frightful shriek. I don't know whether the poor bird or Auntie Fan had the biggest shock.

Jim got his next dog from an old man who told him that it was a one-man dog and might not settle. It did settle and took to Jim but wouldn't have anything to do with anyone else. If Jim was away the dog would wait in the tractor shed, convinced that Jim was doing tractor work. He wouldn't budge; we even had to feed him there.

One time, when a new shepherd came he bought his little mongrel bitch with him. He didn't look after her very well and she was always hungry. One day I heard the baker laughing. 'Watch this!' he said and threw a large roll in her direction. She didn't catch and chew it like a normal dog but swallowed it straight down. I was amazed she didn't choke. In the end she decided to move in with the cowman whose wife fed her well.

She proved to be useful for getting even the most stubborn cattle to move by swinging on their tails so they couldn't kick her. One evening Mr Kneebone wanted some cattle moved and I went with him to get the cowman. He said, 'Bring the dog, Jack,' but Jack shook his head. 'It's no good, she won't come, won't work after half past five.' Mr Kneebone thought he was having his leg pulled so Jack called and called, but despite wagging her tail in friendship she continued to lie outside the back door. The reason why she wouldn't do overtime was very simple. The cowman's wife gave her supper at six o'clock.

One afternoon on the Barton produced the proudest moment of my life, notwithstanding all that I achieved afterwards. After milking in the morning I had a couple of hours to fill before I went to see what inside work was waiting for me. In winter it was attending to the housed cattle and in summer it was checking the non-milking stock.

On this day I came across a steer which I felt unhappy about for no particular reason, just a gut feeling. Endemic to Cornwall was a disease call 'Red Waters'. The name speaks for itself as the specific symptom was dark red urine caused by leaking blood. Early treatment was essential to save the animal so I went back and reported my worry.

Mr Kneebone was going that way in the afternoon and took Jim (the foreman) and me so that we could bring the animal back if he was ill. Mr Kneebone was in his sixties, Jim in his forties, both had a lifetime of experience with cattle and they thought the steer was fine. We stood around waiting for him to have a pee and I got more and more embarrassed, thinking that I must be wasting our time. I was imagining them shrugging their shoulders and never trusting my judgement again when, miracle of miracles, the steer performed and it was dark red! I have never been so chuffed, before or since. I had arrived and could call myself a stockman.

The Barton had two bulls, a huge South Devon called Gerston Doncaster and a much smaller Guernsey. Like most males, when females were not present they got on well together in the field. The Guernsey was the boss and ruled the roost. If handled properly most bulls, especially those of the larger breeds, are quite docile although Jerseys can be quite belligerent and Guernseys are not much better.

We would never have taken liberties with the Guernsey bull but when the coast was clear Lucy (the cowman's assistant) and I would bet each other how much the South Devon weighed. I had the keys to the weighbridge and so we marched him across the yard. The last time we managed this he turned the scales at 23 cwts which is about 1170 kg!

The next time we tried it we weren't aware that Mr Kneebone was home. We heard an upstairs window open and a roar equal to that of the bull. 'Where do you two maids think you're taking that bull?'

I tried to think of an excuse but honest Lucy said, 'Us be going to weigh 'un, maister.'

'Put him back, put him back.' The window crashed closed. That was the end of those little excursions.

It is the sheer size of the animal which can lead to problems; compared to them we are puny, fragile creatures. A bull only has to shake his head and if yours is too close to him you could easily get a fractured skull. If you are close to a wall and he moves towards you, you could easily get crushed.

This was the first farm where I had any dealings with sheep (on the smaller farms grazing could not be spared for sheep because milking cows were much more profitable) and I soon discovered that normal treatments were useless for them. While treatment could usually cure a cow, sheep seemed to have a death wish and would die if they could.

If sheep became sick there wasn't a lot of point in trying to get them well. Instead, we would take them to a suitable place to dig a grave while they were still able to walk and prop them up with straw bales. I talked to them while I was digging the hole although I don't

ɔuth Devons were the usual breed when I worked in Cornwall. They were promoted as 'triple ɹrpose': milk, cream and beef. The cow was photographed in 1948 at Fowey; the owner was ʼrs K Hawke of Hillhay. The bull was the winner of the South Devon class at the Liskeard ɟricultural show in 1949.

ɔrnish Studies Library E16766 and E17907)

I learned to love livestock on my grandparents' smallholding. Goats and their kids were always inquisitive and entertaining. We taught hens to ride on our laps on the swing, though I never understood why my grandparents kept so many chickens.

he old-fashioned gipsy caravan that my mother bought. I'm sitting at the top, ruling the roost ver two of the neighbour's children.

Walling was one of the jobs that took my fancy. I had a go at it - but perhaps not quite as expertly as these men at a young farmers' event, Liskeard, 1950.

(Cornish Studies Library E19936)

Horses being brought in for shoeing at Washaway in 1949. Taking the horse to the blacksmith was one of the first jobs I was given in Cornwall, to keep me out of mischief while the farmer went to market.

(Cornish Studies Library E17764)

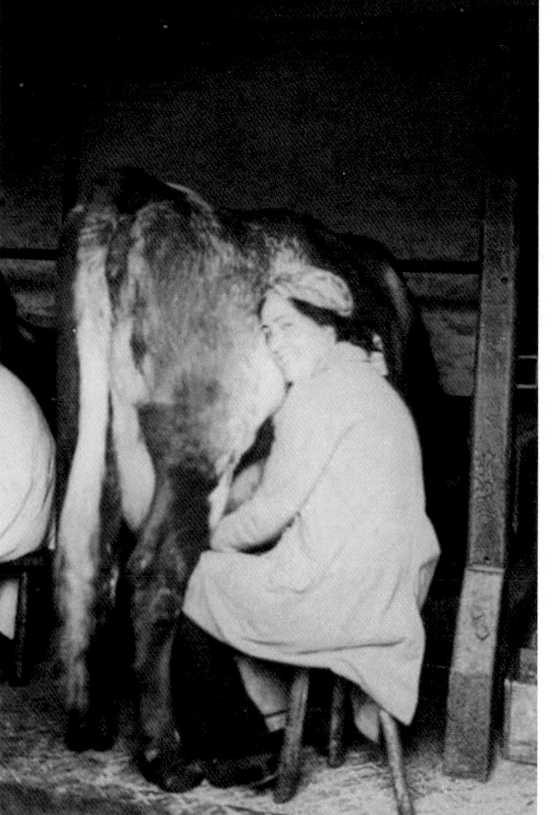

There were two Bubblewell farms near Minchinhampton. This (above) is much the bigg of the two. I lived at nearby Lower Bubblewell farm from 1940, starting full-time work there in 1942. The feeling of spaciousness in the rich pastures and commons on the high Cotswold plateau is recalled by this postcard from thirty years earlier.

(Minchinhampton Parish Council)

I dread to think how many cov I hand-milked in the 1940s an early 50s. This picture (left) is believed to be of Marjorie Cripps working at Upper Bubblewell Farm.

(Minchinhampton Parish Council)

It took me a lot of perseverance, but in the end I finally got to do what I most wanted – take charge of tractors, including a Fordson.

(Royal Institution of Cornwall, Courtney collection AG448)

ıst before the Land Army as disbanded in 1950 I was ılected as one of the girls to present Cornwall at the ıtional winding-up parade. night be in this group elow), among the five ındred girls marching into ıckingham Palace, not ımpletely in step.

'useum of English Rural Life)

I didn't have much to do with sheep until I went to work in Cornwall, then it was mainly with South Devon longwools. George Ellis's 1948 photograph is of champions at Nanswhyden, St Columb Major. (Cornish Studies Library E17764)

I met some remarkable farming characters when I worked at Caerhays. The castle stands on the estate in an idyllic location protected by woodland and with pastures sloping down to a sandy cove. (Roger Smith)

know whether they appreciated it or not. I made them as comfortable as I could until they died and then took away the bales and rolled them into the hole.

The shepherd was a tubby little man and I was encouraged to accompany him on his lambing rounds because the boss thought that his heart was a bit dicky and he should not be on his own in the cold and wet. One really frosty night he remarked, 'I shouldn't like to be out here in just my socks.' I had a wonderful vision of this gnome-like little man with the moon shining on his bare buttocks.

In those days we used to lamb outside in February. The in-lamb ewes were put into small fields near the buildings. Of course, they chose to lamb at the most inconvenient time for us; usually the middle of the night. It was not the custom to stay up with them all night, instead, we went around at about 11 o'clock with a torch with a good beam. I often thought how difficult it must have been before these torches became available and shepherds had to go around with a hurricane lamp.

We inspected any sheep which were lambing in case they needed any help although they usually managed quite well by themselves so long as the lamb was presented correctly. If the lamb came backwards (a breech birth) we had to act quickly since it meant that it was likely to be short of oxygen. Sometimes the head or a leg would be bent backwards so we had to push the lamb back in order to bring the appropriate body part forward. Have you ever tried to find something in a drawer with your eyes shut?

Unlike upland sheep which must deal with harsh conditions, cosseted lowland sheep have never had to develop a good brain which makes for a real problem when they are required to count. If a ewe had a single lamb she would lick and fuss it and it would get to its feet and suckle. However, if the ewe had twins, after having the first one she would often move some way away to have the other and leave the first behind, completely forgetting about it. We needed to be there to follow her around with the first lamb to make sure she suckled them both and they would be safely bonded.

Fortunately, the shepherd was very good at knowing which ewes where likely to be carrying twins and we would come out early in the morning, hoping we wouldn't find any spare lambs. If we did we would try and guess which ewe it belonged to, otherwise we had to try and persuade a ewe who had lost her lamb to adopt it.

If we wanted a ewe to adopt a strange lamb we put them into a small pen together. Generally the ewe would try to kick and butt the lamb when it tried to suckle. It quickly learned not to when we *were* there. We usually had to climb into the pen and hold the ewe.

After a couple of times they learned how to behave and all we had to do was look at them. The lamb rushed in to grab a teat and the ewe stood stock still glaring at us, obviously furious.

Why, when you cast amorous eyes at someone, it is referred to as sheep's eyes I can't imagine. They have cold yellow eyes which is hardly the way to look at someone you want to impress. Cow's eyes would be a better metaphor. They have bright, warm eyes with curved lashes better than a model could find in her make-up box.

Once the ewe allowed her adopted lamb to suckle when they were in the pen, we turned them out into a small paddock with the ewe on a long rope. We checked them often and the rope meant we could stop the ewe running away so the lamb could drink. After a few days the ewe accepted her adopted child and they were turned out with the rest of the flock.

'Turned out with the rest of the flock', sounds easy doesn't it? Moving a single ewe and lamb can often have complications. You start off in the idyllic Christmas card manner with lamb tenderly tucked under your arm and the trusting ewe trotting beside you. This might work. With luck, ewe and lamb will be delivered to join the rest of the flock.

However, not only can ewes not count but their memory is as bad as an advanced sufferer of senile dementia. The ewe would abruptly stop dead with the attitude of a matronly lady to whom it suddenly occurs that she has left the gas on. She would turn around and set off the way she had come, thinking, 'Where's my baby, I've lost my baby. Where's my baby, I know I left it somewhere.'

Holding out the lamb and making appropriate bleating noises was useless and so you had to go all the way back, carrying the lamb, to capture its mother. I put something round her neck; usually my belt, and then I could drag the protesting ewe to where we wanted her.

Lambs were never counted until after lambing was finished. It was considered unlucky to count before. I should imagine that this was because if you told the boss how many lambs you had but lost some then he would want to know why. It was much wiser not to be able to tell him.

Rams had a much more laid-back existence. Most of the year they led a bachelor life and kept to themselves in the manner of gentlemen in a London club, well away from the screaming infants they had sired. When all the sheep were turned out together these old fellows would cluster together well away from the ewes and lambs. Of course, this was quite different from tupping time in autumn when they had to be kept well apart.

Rams were put in with the ewes in autumn and then would be raddled in order to keep track of which ewes had been served. Raddling was normally done by putting a special harness on the ram but our system was much more basic. We rubbed his chest with different colours, starting with red lead, which was used a lot in those days, and working onto darker shades. The ewes were daubed with the same colours so we could predict when each of them would lamb.

When a young, almost certainly urban, teacher came to the village school she addressed the eager, up-turned faces, 'Now children, I want you to look around and observe everything that goes on. For example, my husband and I were walking in the Pound field near the cliff path when we saw that the sheep were marked with different colours, ready for when they go to market.'

'Please, Miss,' said the shepherd's son, who was all of eight years old. 'That's not for market, it's because the ram's been on them.'

A dog can be useful with cattle but sheep are harder to manage and there are not many people who try to cope without a four-legged helper. Normally Welsh or border collies were used but I worked with several dogs of unspecified ancestry who were quite adequate.

I heard an amusing story about a collie who would be sent out to bring in the cows by himself but sometimes absent-mindedly brought in the sheep by mistake. His owners said he realised what he had done and would go and hide until he thought it was safe to come out.

In Cornwall we never worked a dog out of sight because if something went wrong you would never know and if a dog got upset he might nip the sheep. I never saw it used but was told that a good way to stop an over-eager dog was to remove the shot from a shotgun cartridge and replace it with wheat. If the dog continued to harass the sheep after you told him to stop, you shot him with the wheat-filled cartridge. This would give him a good sting but the sheep's woolly coats would stop it hurting them.

Chapter 9

New Skills and the Urban Outlook

WHILE I was at the Barton I managed to have a go at most of the jobs that I had always wanted to try.

I had been keen to have a real go at stone walling ever since my first attempt in Gloucestershire. It was thanks to a minor disaster which forced me to do some repairs. I was chain harrowing with the horse and misjudged the gate from one field to the next so caught the corner on the way through. The mare felt the resistance so put all her weight into her collar and brought the whole structure crashing down. I had to set to work and build it up again.

In Gloucestershire the stones are more like bricks and the odd shaped ones which we called 'horses heads' were put in the middle. Walls were built completely with stone.

In Cornwall there is less stone which is easy to get at and it doesn't break up into nice evenly shaped pieces so you have to learn to use whatever you have. Only the outside of the wall is made of stone and the middle is filled with earth. In fact, in some parts of North Cornwall they became earth banks instead of walls. It was quicker and easier than stone walling but you could just hear the rabbits sitting and saying how kind we were to build them such a nice dry home.

Unlike the horizontally laid stones in the Cotswolds, in South Cornwall each stone was placed vertically, either straight up or in a herringbone pattern for decoration. I only ever tried the ordinary straight-up method. Walls were not vertical from top to bottom, but dropped back a few inches for every foot in height. The amount of fall-back decreased with the height of the wall. The result was a firm-standing wall less likely to bulge in the middle. From a sheep's perspective the walls appeared to have an overhang which they were less likely to try and jump.

Jim found me a broken piece of wall and a load of stones and I borrowed a hammer from the shepherd (it was normally his job to keep the walls in good order). I was slow so practised after work. One evening I heard a noise behind me and it was Mr Kneebone, puffing and blowing, on the little black pony he rode. He said, 'Pull it down, pull it down,' and left.

When I returned home I told Jim, 'He came' (we all knew who 'he' was) 'and told me to pull it down.' Jim replied, 'Oh, you needn't worry, he made me pull down the first one I did six times.'

When I got to the office the next morning Mr Kneebone was there with a beautiful drawing of how I should do the wall. When I arrived that night he was waiting for me. He could see where each stone went so would get impatient while I looked around for the right one and poke me and the stone with the stick. In the end I did a good enough job to satisfy him, and as far as I know the wall is still standing.

When I was working at the Barton I was also allowed to have a go at shearing although, again, I was only allowed to practise in my own time because I was very slow and you can't rush when you are learning in case you cut the sheep. Shearing sheep and ironing clothes are very similar, you have to get the wrinkles out before you start.

We would open up the sheep down the middle while it was sitting on its bottom and then shear right round one way, change hands and shear the other way. The wool dropped off all the way down and then you just had to do a little bit around the back legs and bottom of the back.

Lambs were harder to shear because they wriggled and kicked. With many breeds of sheep you do not sheer the lambs but these were South Devons which are a long-wool breed which produce quite a lot in the first few months of life. We didn't get fleeces, just loose wool, but in the days just after the war we could sell anything.

The essential thing to do before you started was divide the flock by sex so that you did not have to stop to check each one. However, since it was a small group of lambs I decided that there was no point. I had one terrible shock when I had a ram lamb instead of a ewe and thought I had cut off his essential equipment. Fortunately I hadn't. I don't know what I would have done if I had.

The other job that I had really wanted to have a go at was thatching the ricks. Of course, this is very different from thatching a house. The straw is laid flatter and you use a lot less. We always tried to do a beautiful job and it seemed such a shame that the rick only stood for a few weeks.

Before being let loose on a rick I practised on mangold clamps which have to mature until after Christmas, otherwise they will make the animals ill. In most parts of the country they were covered with straw and then earth to protect them from the frost but winters in Cornwall are warmer and wetter so a different method was used. First they had a coat of rough straw, then hedge parings such as brambles were used to break the frost. Finally, the thatching kept out the rain.

Every area where I worked had a different method of building ricks. Sometimes they were circular which must have been very difficult. Other parts of the country built the ridge shorter than the eaves so that all four sides could be thatched. Cornwall, of course, was different. The ridge was longer than the eaves so that each gable end overhung the bottom of the rick and it was only possible to thatch the two long sides.

Before starting, a 'dolly' was made which was a long bundle of straw the same length as the ridge. We fixed it to the top to give a good point to the roof. Next, four thick bundles were fixed to the gable ends which covered them once the rest of the rick was thatched. Thatch was held on with string attached to pointed sticks. These 'spekes' were about eighteen inches long and driven into the rick. We had to be careful that they did not form rows from the top to the bottom of the roof otherwise it made a runway for the rain to get in. Instead they were used in the same pattern as bricks are laid.

The Barton had its own threshing machine and, unusually for the time, enough people to run it, so over the years that I was there I managed to have a go at each of the jobs on the machine.

If I was responsible for filling the sacks with grain I was worried that if I did not fill them fairly full the men would think I was insulting them. In the end it turned out that they thought that I was trying to kill them since a full sack of wheat weighed about 18 stones (127kg).

The worst job was the final one which we called 'raking doust'. This meant keeping the underneath of the machine clear of debris and I always got absolutely filthy!

Threshing days were the only ones when we had a break in the afternoon. At about four o'clock we were sent out a very welcome can of cocoa. I always found three until four the hardest. After that I seemed to get my second wind and then I could have gone on all night.

I do not know what idea urban people have about farm workers nowadays. They probably don't even think about their existence but way back in the 1940s urban people looked down on country people as being slow in movement and thought. There was nowhere that I learned more about the importance of slow movement and forethought than at the Barton.

In the city the saying goes: 'A fool and his money are soon parted.' Urban people didn't realise that in the country a fool is soon parted from life and limb. We worked with animals that meant us no harm but were bigger and heavier than us and didn't try to be careful. We had to concentrate. If we acted without thinking we could end up in serious trouble.

It was vital to let an animal know if you were approaching. Otherwise they would get a fright and kick, bite or run away and knock you over. One man at the Barton who got caught out like this should have known better. He had been working with cattle all his life but never thought what he was doing.

He was a heavy chap and, like some heavy people, could move very quietly. On a very wet February day he came into the cowshed to shelter. The shed housed a few Guernsey cattle who had their heads in the manger because they had just been fed. He came silently up behind them and when one of the cows saw his shape behind her she lashed out and caught him on the shin.

If this had happened to a lightweight person, like I was in those days, they would just have gone flying but this man's weight meant that the bone snapped. The break never really healed and I believe that in the end the toxins got into his blood and killed him.

We had a field which we referred to as the moorland. It was too wet for the tractor to get onto to clear out the ditches and so men who did not have much else to do would clear them by hand. One day while the careless man was doing this job the foreman and I passed him on our way to get some cattle and stopped to talk. He was a hard worker and continued digging while he talked but his coat was rolled up by the side of the ditch and the foreman gave me a nudge so I saw what was going to happen. Sure enough, he carried on shoveling out the black gunge and buried his coat with it. This man had gone on for years like this but eventually lost his life through not thinking about what he was doing.

Another reason for moving slowly and gently was to avoid stress or upset to the animals. They are much easier to handle that way. Exercise is good for cows but it does them no good to be chivvied and rushed. Apart from anything else, you don't run a marathon at the speed that you sprint for a train and we had to be able to keep up a steady pace all day.

Urban people also had the idea that farm workers had no manners, although in reality we just had a different set of rules and etiquette for looking after each other.

For example, if you were the one driving the horse or tractor which pulled the trailer that someone was loading, you had to remember to call, 'Hold tight' before you moved off, otherwise they might lose their balance and fall.

Another unwritten but vital rule was that if you were pitching something up such as sheaves of corn and the loader failed to catch one, you did not put up your pike or pitch fork to catch it. Instead, you let it fall and then pitched again.

The reason for this was to prevent what happened to me. I was working with a fellow who tried to catch a falling sheaf and the fork went into my hand. Fortunately, the fork was clean but there were no tetanus jabs in those days so I was worried. I was careful to put plenty of antiseptic on the wound and maybe that did the trick because it was alright.

The same man told me that he had been doing some circular sawing with another Land Girl and taken her fingers off! After that I always made sure that he did the sawing when we worked together, even though it was less strenuous than manhandling the branches which we were making into logs. We were never allowed to talk when we were doing circular sawing.

I am sure the urban population would have been surprised to know that they were thought of as complete idiots by country people. One of the farm workers whom I met in Cornwall told me about a fisherman who lived there before the war when visitors came to the area. One such visitor greeted the old man with, 'Good morning. Oh, what a beautiful day.' I would have been as ignorant as her about fishing matters but apparently the weather was completely unsuitable for the old man to go out fishing and he thought that everyone should know this. He replied, 'You'm too foolish for me to speak to.'

I expect they thought me a useless townie as well until I managed to prove otherwise.

Just before the Land Army was disbanded in 1950, and while I was working at the Barton, there was a parade of five hundred girls to be inspected by Queen Elizabeth (the wife of King George VI) at Buckingham Palace. There were a few girls representing each county and I was selected for Cornwall.

This was the first time I had travelled by sleeper on the railway and when I arrived in London we were taken, I think, to Clapham Common. I was surprised to discover that there was a vast accommodation complex under the common, presumably for forces during the war.

The following morning we were loaded into some twenty-eight buses, about six hundred of us including reserves, and taken to The Mews to be drilled by a Guard Sergeant Major.

When they told us to form three ranks there was a great deal of shuffling and pushing because no-one wanted to be at the front. We gradually retreated across the parade ground until someone took charge and lined us up; then they proceeded to try and drill us.

None, or at least only a few, of the girls had been drilled before and the sergeant gave his orders in standard British army fashion which can only be understood by soldiers who have been drilled. It was total chaos.

The next morning we were lined up again and inspected by the Land Army authorities to make sure we were properly turned out. They even measured how far our socks were turned down! Then we set off on a march to Buckingham Palace accompanied by two bands, one of which was the Irish Guards, and went through the streets of London, one, two, three, hop. If you decided that you were out of step, at exactly the same time the girl marching next to you would decide she was as well. Then you were back to square one.

When we reached Buckingham Palace we formed an open-fronted rectangle with the bands behind us. A few yards in front of us were the girls who were to receive long-service awards. The Queen inspected us and then stood on a rostrum to present the medals. So ended my time in the Land Army.

Leaving the Land Army made little difference to me since I continued to work at the Barton, except that I now had extra freedom to do as I wished. The disadvantage was that there was no more uniform and so once it wore out I would have to buy my own clothes. Fortunately, I had managed to save quite a bit; we had run a club for old uniform which we would share and present to the authorities to prove we needed new. I am sure they became familiar with some of the garments that were passed around but they were very good about it. I was only once asked if I was sure that the coat I was holding was mine.

Chapter 10

Two Remarkable Men

THERE were two remarkable men at Caerhays. The first was Mr Kneebone, the manager, who reminded me of an overgrown hamster because of his rather red, pouchy cheeks and protruding front teeth.

He was a portly gentleman, about sixty-five years old who weighed some seventeen stone. His wife put 'V' patches into his trousers but she never seemed to be able to match the colour exactly so he would either be in dark trousers with a light 'V' or light trousers with a dark 'V'. He showed me his fingernails once which were blue and told me it was obvious he had a heart problem, although he went on to live into his eighties.

He had an elderly dog, a rough-haired terrier which would go to sleep under the table where I was working. If it was warm in the afternoon both the dog and Mr Kneebone would go to sleep and snore. He would wake up, blink owlishly and say, 'Was I asleep?' He knew in his heart that he was past it but he had given his whole life to the farm and estate and couldn't bear to part with it.

Several years later he was forced to retire to a bungalow in St Austell and I went to visit him. He took me into the garden and said, 'Look, you can see the Barton from here.'

He and his wife had good senses of humour which they kept well hidden. Mrs Kneebone would say that she had seen old Alec sneaking away for a quiet smoke so Mr Kneebone would give him ten minutes in peace then creep up and 'find' him.

He was a man I respected because he knew exactly what he was talking about. He had learned all the practical skills when he was young and so when one of the chaps couldn't get something right he would say, 'Get off there, I'll do it,' and he would show them how to do it correctly.

Mr Kneebone had come to the Barton in 1910 when he was twenty-five and had lodged with the shepherd's wife whom I came to know well. She was not a vindictive person but said to me once, 'Why is it that he who arrived with next to nothing should now have all this respect and high living while my husband is still struggling on the low wages he is offered.' It was because Mr Kneebone had had thorough training in practical skills and farm management and, most importantly, he was prepared to take responsibility which was the last thing that most of the other competent farm men wanted to do.

Of course, this gave him an overblown opinion of himself, and the foreman often told me that he was ashamed to go to the market in Truro with him. They would go into Farmer's Factors where Mr Kneebone would march to the front of the queue and bang on the counter with his stick. An assistant would come rushing out immediately to see to his requirements. To him it was a perfectly natural way to behave.

I got on with him very well but I realised very quickly that he was not someone you could answer back. The foreman, Jim, told me about a time several years before when they were manoeuvring the threshing machine by hand since it had to be completely level to work properly. One of the men was not trying very hard and so Mr Kneebone said, 'Push Robinson, Push.' Robinson replied, 'Push it yourself, Mr Kneebone.' Straight away, Mr Kneebone said, 'Cards, out.' I thought this was a bit harsh but Jim explained that after the First World War the estate had taken on no end of unemployed workers and without this attitude he would have lost control.

Only once did he exasperate me so much that I did answer back. Fortunately, there was no-one else around at the time otherwise I would have been in trouble. There were about half a dozen bullocks, over half a ton each with Mr Kneebone behind them. You would try and get away if Mr Kneebone was behind you as well but unfortunately they were trying to get through the gate that I was trying to close. Mr Kneebone was shouting, 'Shut the gate! Shut the gate!'

So I replied, 'What the hell do you think I'm trying to do?' As soon as I said it I could have bitten off my tongue but he took no notice. Perhaps he realised it was his fault.

He could move very quietly and one of his little tricks was to creep up on you unawares. Sometimes he would come in when we were milking so we had to be a bit careful what we were saying.

When we were building a corn rick I used to put in the middles and it amused me greatly when Mr Kneebone's head would suddenly appear with no warning and say, 'That's right, that's right, keep her belly full.' He said this because the middle of the rick always had to be higher than the outsides so that rain would drain off rather than sink to the middle and cause rot.

The greatest source of hilarity was when we were picking out cattle ready for market. Four or five of us would stand about to keep them close. He would wander about, prodding at them to choose the ones for market and say to us, 'Keep still, keep still, you'll frighten the bullocks.' Of course, we were just standing there so the only one frightening the bullocks was him. The others were not so amused but I used to think this was marvellous.

The only real irritation was the way he changed his mind. He would send me out to tell one of the men to do something and then half an hour later he would tell them to do something else. Maybe it was because his memory was bad but when I said, 'I do wish you would make up your mind,' he would reply, 'If you keep on changing your mind you'll never go mad.'

I really got a lot of enjoyment out of working for him. He would often say, 'Are you doing anything this evening?' He knew full well there was nothing to do. We were eleven miles from St Austell, fifteen from Truro and I only had a bicycle. There were two buses a week which went directly from the village, one for shopping on a Tuesday afternoon and one early on Saturday evening which was known as the picture bus. You had half an hour to do your shopping. Then you could get to the pictures and come back again at about 9.30. For any other public transport you had to walk five or six miles to Gorran Highlanes to get a bus to either Truro or St Austell.

Mr Kneebone's interest in my evening activities was that he wanted me to go out with him in the van to open gates while he drove around the farm. It was a two-wheel drive van, nothing to get you out of trouble and a lot of the fields sloped towards cliffs which were basically unfenced. He would be busy looking around at things and although I learned a lot from him this way

I was in terror most of the time, with the thought that at any moment, if the grass was slippery, he would lose control and we would go over the cliffs. I always sat in the passenger seat with the door partly open. Just in case I had to jump out.

We had another 200 acres near Truro and on Boxing Days he would take me there in the van because the whole place was riddled with rabbits. He would sit on his shooting stick with his twelve bore. I would put the ferrets in the holes and when the rabbits came out he would bowl them over.

The roads in Truro were narrow and twisty but fortunately, at that time of year, there were not many visitors about. One day he said to me, 'Someone told me that if you drive straight across the corners you save an awful lot of wear and tear on your tyres,' and he proceeded to do just that. Blind corners! If someone coming the other way had crashed into him he would have said, 'What were they doing there?' Driving with Mr Kneebone was a lesson in controlling your nerves.

The second remarkable man at Caerhays was Jim, the foreman. He had been trained by Mr Kneebone as his right-hand man so he shared his outlook.

Each morning we would start milking at 7.00 and at exactly 8.00 Jim would climb into the enclosure beneath Mr Kneebone's window which contained a few ducks destined for the table. Since Mr Kneebone would still be in bed, his wife would come to the window and relay messages to and from Jim. One of the estate workers came into the yard and commented, 'Eight o'clock, holy communion.' Jim would then come round and give everyone their orders for the day. This included me if I didn't have much to do in the office. Jim could have run the farm quite well on his own but major policies came from Mr Kneebone.

At the time I didn't appreciate what an incredible man Jim was. He was another large man, probably as big as Mr Kneebone, but not as portly. It was all muscle, no fat. But they weren't pretty muscles. He wasn't pretty in any way, in fact, quite the opposite but there was no job on the farm which he couldn't do better than everybody else. If you ever had a problem you took it straight to Jim.

Jim would undertake difficult ploughing and the drilling. In those days you had to have someone standing on the back of the drill to see that the seed was going through and stop it operating on the headlands. This was often my job and I would watch as he lined up the radiator cap to a point on a far distant hedge. The first line would be perfectly straight and the rest would follow.

He was also accomplished at rick thatching although his method of bashing down the sheaves to give a flat surface would frighten me. He would lean back and forth on the long wooden ladders that we had in those days and it was an extraordinary sight to see this great big man sway back to the point where if he had gone any further he would have fallen over backwards and smashed himself to pieces.

He could also recognise problems with cattle and sheep and shear faster than anyone else on the farm – about eighty sheep a day. New Zealander and Australian farmers would think this very few but these were large South Devon sheep with thick, heavy fleeces. We once had a big ram which two of us held while he was being sheared. The fleece weighed 36 lb. For Australian chaps shearing is their specialised job, for Jim it was once a year.

But for all his skills, the thing that impressed me most about him, thinking about it afterwards, was his sensitivity. I remember coming back from loading up the fat cattle for market when he said to me something he would never have said to the other men: 'I hate that job, I feel like I'm betraying my family.'

Many people think that farmers and farm workers are hard but we genuinely loved these animals and gave them the best life that we could. We lived with them, worked with them and loved them.

Another time Jim and I were walking through the woodland just before one of the first open days on the estate and he said, 'This is paradise. Tomorrow it may be heaving with people but today this is paradise.'

The most astonishing achievement and ultimate indication of his sensitivity was one day in the summer when he came home very upset. He had been at Portluney Beach and seen a family with their little short-legged Scottie-type dog. They tied the dog to the rear of the car because it was too hot to leave him in it or to take him to the beach. When they returned to the car they had forgotten all about the dog and joyfully drove away, heading for a very steep hill as they left the car-park.

Jim was a stockman and would never have forgotten about an animal. He shouted at them but they didn't hear him so he set off on a route different from the one they were taking. It was flatter ground but about a mile in length and he was not built for running; in fact, I don't think I ever saw him run.

He arrived at the top lodge in time to meet the car. 'I stood in the road,' he told us, 'to make them stop. I got out my knife and said, "Don't you dare move." They probably thought they were going to be robbed by some idiot countryman. I went round the back and cut off what was left of that poor little dog and dumped it on the man's lap. "Now you get to the vet," I told them, and they all burst into tears. I was so furious I didn't dare stop and walked away.'

That was Jim. The effort he must have put into running that distance is quite extraordinary.

I never realised at the time what a wonderful man he was because I was going through my feminist period and he needled me at every opportunity. Like a fool I rose to every challenge, so we had quite a few heated discussions.

When I first moved to the Barton I went to live with Jim and his family. It was a three-bedroom house and Auntie Fan was living with them so when the boy needed a room I had to move out. They found me a little cottage, one up, one down, a few yards from the sea wall at Portholland. I was lent some furniture from the castle but still had my main meals with Jim and his family.

The cottage was in a very interesting position. If there was a high wind or the tide was in then the sea would come right over the wall. This made getting in very difficult. I waited for a wave then ran forward, put the key in the lock and dived out of the way. After the next wave I turned the key and took it out. After the third wave I could just get in before I was drowned.

Once I was inside, the cottage had been so well built that I hardly knew that anything was going on outside. It was a really nice place. There was no bathroom but there weren't any anywhere. This wasn't because the estate wouldn't put bathrooms in but, as they explained, there was no water.

The water came from a stream that ran into the sea at Portholland. It went through a little filter bed then was pumped up a hill of one-in-three to Caerhays by a hydraulic ram. Nowadays you would never be allowed to use the water because higher up the hill, before it reached the filter bed, there were two farms where water used for washing the cow sheds could get into the stream. None of us suffered any ill effects from it.

The problem with this cottage was that it had just a stone floor. Also, the tenants before me had asked for the Cornish range to be exchanged for an open fire. This was foolish since a Cornish range was a good cooker.

Winter was approaching so I asked if I would be able to keep the cottage and make some improvements. The agent at the head office told me that they would probably need it for Alec, the handle-turning general worker. From my point of view this was pretty insulting because it seemed like they put more value on Alec than on me. I wouldn't have minded if I'd been offered an alternative.

I thought about what I was going to do. I was still in my early twenties and keen to learn. I wasn't going to be able to do any tractor work at the Barton since we had two tractor drivers and I couldn't take work away from them. I decided in that case I had better move.

When I started looking through the *Farmer's Weekly* for jobs I saw an advert for a pupil on a farm called

Scadgill in north Cornwall, near Bude. I didn't have the savings to pay like they used to in the old days but thought that if I was willing to work for very little I might get to learn tractor driving and more field work, rather than being tied to cows' tails all the time. I took off on my newly purchased two-stroke 125 cc motorbike in my smartest uniform, although, I suppose I shouldn't call it a uniform since I was no longer in the Land Army.

After the owners interviewed me they said, 'We're ever so sorry but we couldn't possibly take you as a pupil because you've been in farming longer than we have. We would be quite happy to take you as a worker, though, and you can definitely do some tractor work.'

So that was it. I went back to the Barton and handed in my notice. It was nothing to do with the people and I had really enjoyed my time there. I was just annoyed that I was so under-valued that I wasn't even worthy of the smallest cottage.

If I had been a bit older I would probably have gone to Mr Kneebone to tell him why I was thinking of leaving and I'm sure he would have sorted something out. But, of course, I didn't. So I set out for my new home.

Chapter 11

Mud Highlanders

THE people whom I went to work for near Bude had only recently come into farming. They had returned from Newfoundland – where the husband had been the manager of Gander airport – because they had wanted their two sons educated in England. He was the first man whom I called 'Sir' and, although I found it very embarrassing, I didn't know what else to call him because he was a Wing Commander. 'Mister' or 'Boss' didn't seem to suit somebody of his rank and I couldn't call him by his Christian name, so it had to be 'Sir'.

The farm was about 200 acres and apart from the bit where the farmhouse stood, most of it was very steep. If you had knocked the tractor out of gear you could have trundled all the way to the sea if it weren't for the walls and ditches in the way.

Most of the soil was shallow, perhaps two or three inches. This was very frustrating when I was trying to put up electric fencing. On the other hand, it was also very wet and boggy with lots of shale. Not the sort of land you would buy if you could afford to be choosy about it. Even in those days you needed a considerable amount of capital to buy a farm so this family had chosen somewhere where they could afford a reasonable size farm, albeit not with the best soil.

The boggy land on this farm has given this chapter its name. We used to put sacks around ourselves to keep some of the mud off our clothes and, if it was raining, one over our heads as well. One day someone came out of the house and said, 'Oh, mud highlanders'.

It was a mixed farm although the main enterprise was the dairy. They had bought the Ayrshires as young stock because they had relatives in Scotland who could obtain them. Then they had to wait until the heifers calved down to get some income. There were also a few sheep, a few pigs, six hundred chickens on deep litter and hundreds of wild rabbits eating what was meant for other mouths.

I lived in with the family. It was a very old house, seventeenth century I believe, and built with thick walls of the traditional soft earth material known in those parts as cob.

The winds that blow on the North Cornish coast can be quite dramatic and even on a calm day you could tell what it would be like because the bushes on the tops of the walls all grew sideways. As I was lying in bed a short time after I arrived, I thought that the bed was swaying but told myself I was imagining it. In fact, the whole house was swaying in the wind!

The house had electricity, just enough to run dim lights. It was generated by an old engine and stored in submarine batteries. We had to keep an eye on the lights the whole time because if they dimmed it meant that the engine had stopped and the batteries would be driving it backwards. We would have to run out to the generating shed and switch it off before the batteries ran out of charge.

Over the years mice had burrowed their way through the walls and by nine o'clock at night you would hear a rustle by the waste-paper basket and a mouse would be sitting by the Aga cleaning its whiskers. These mice provided sport for us in our attempts to catch them.

Having placed a bowl of water in the sink we would turn off the lights and wait for a few minutes. One of us would say 'one, two, three,' and turn the lights back on so that the mice would scuttle along the pipes over the sink to get home. The other would charge into the kitchen with a little stick in their hand and knock the mice into the bowl of water. The mice would swim around until we tipped them out in the yard.

In the end the family had to call in the pest controllers who got rid of them for a lot of money. Unfortunately, they could not get rid of the rats because they would just come back. They could be a bit off-putting if you went onto the rick yard at night. If one had got onto the rick it could get frightened then jump out and land on you. Fortunately, I never had this happen to me!

The family had a dog called Chewby. He was originally called Toby but when he was young he had chewed everything and so his name was changed. He was a Labrador cross and not a very bright individual. He used to love to chase the rats in the barn but if he managed to catch one, instead of killing it and putting it down, he would run around and try to catch the others with the first one still in his mouth.

A trapper came for the wild rabbits every winter and the proceeds were split between the farmer and the trapper. One year I remember the farm receiving £200 which was a lot of money in those days, almost a man's wages for a year.

The trapper had a little old pony with dozens of gin traps slung across its back. We told him that when he got to an electric fence he was to look for the red box to switch off the current and then he could unhook or hold up the wire for the pony. He wouldn't do that, instead he would hold up the wire with a stick so he didn't get a shock and let the pony walk underneath. One day he didn't hold it high enough and one of the gin traps caught the wire. You can imagine the shock that the poor pony got. We could hear the rabbit trapper cursing and swearing for miles and it took him half a day to catch the pony. Whether he ever got it near an electric fence again I don't know.

We used the electric fencing in the summertime to control grazing. We back-fenced which meant that once the animals grazed one area we put the fence across to stop them getting back onto it until the grass had grown again. I took great interest in arranging the fences because wherever the cows were, they had to get back to the drinking trough. We didn't carry kale to the cows in winter but let them graze it and used the electric fencing to ration the amount they had.

The milking was done by a lad who lived in the house as well. There were three of them at different times while I was there. The second one of these was a very peculiar chap who had worked as a herdsman for the well-known entrepreneurial farmer called Rex Paterson who rented several hundred acres in Hampshire. The cows on this farm were scattered about in small units and milked in Hosier bails, each group having its own herdsman.

This man made out that he was very intelligent and told me once that he had letters after his name. I should have said, 'Oh yeah, BF I suppose,' but his superior attitude was unpleasant enough without being antagonised. It was quite obvious when he came that he did not know much about dairy work because he began to scrub one of the milking buckets with Vim which you never do. For good hygiene, the last thing you want to do with milking buckets is scratch them.

He got in with a local farmer whom he apparently went bell ringing with, and told us how he was always being asked round for a meal. I met this farmer a while later and when I informed him that this chap had left he replied, 'Thank god for that, I couldn't stand the man.'

A relative of the farm family sent a pony as a gift for the two boys who were about ten and eleven years old. Since I had done a bit of riding I was allocated the job of giving them a few basic lessons on a lead rein but they were not very interested and soon returned to their train sets.

The rest of us took turns to ride this pony and it turned out to be quite a terror. It got most people off. One day our cocky cowman thought he would take it for a ride but when he came back he was very quiet and never offered to take it out again. I am sure that it had had him off and he wouldn't admit to the fact.

The real pièce de résistance was after he had left, leaving behind a quantity of magazines to be inherited by another worker. A few days later this worker came in with a big grin on his face. 'Look what I found among those magazines,' he said. It turned out that the cowman who had been passing himself off as single was not only married but had a daughter, and amongst the magazines was a letter from his wife. 'Our little girl is growing fast and is going to be as clever as you are,' she wrote. It was obvious that he had told her that he was managing the farm. Poor lady, I wonder if she ever found out the truth.

The next guy we had was a student doing his practical year before he went to university and he was completely the opposite of the other chap. The cows found out that they could boss him about and took advantage of it.

We used to put silage outside the shed to be fed after milking, otherwise the smell would taint the milk. The cows had to walk past the silage to be tied up so someone had to stand guard as they came by. If it was Pat they would just put their heads down and push him gently out of the way. Then they could do whatever they liked.

We took it in turns to get the cows tied up even though Pat did most of the milking and someone would have to help him tie up, otherwise there would be bedlam. There was one very smart lead cow who would always get into someone else's stall first and grab the titbits that we put there to tempt the cows to be tied up.

I used to stand at the door and roar at them and they knew perfectly well that they were doing the wrong thing so would stand to attention in their proper stalls. If no-one was there to help Pat the cows would be looking for whatever mischief they could get up to next.

I was always suspicious that when he milked them he got more milk than I did but I was careful not to mention this!

On all the farms where I had worked previously we had hand milked the cows but here it was done with a machine. The machine was designed to allow the operator to milk several cows at a time and required constant attention. The vacuum lead was attached to a line by the cow's head and the milk was collected in closed buckets which we could then empty into the cooler. In those days there was no automatic cut off and so we had to be very careful not to leave the machine on for too long, otherwise it would creep up the teats and damage them.

One person would do the milking while the other was responsible for washing and preparing the cows, taking and weighing the milk, putting a lid on the next bucket and making sure that the churns were changed before they ran over.

Surprisingly, the machine is a lot better for the cows because hand milking means putting pressure on the outside whereas the machine imitates a calf suckling and creates a partial vacuum which helps the milk to flow.

We had about thirty cows so the job took about an hour and we concentrated very hard all the time. There was no time to chat while we were doing it – as there was with hand milking – so it was not nearly as relaxing a chore.

I missed hand milking. It can be quite pleasant and peaceful apart from the kickers once your arm muscles have got used to it. Since each cow can take up to ten minutes to milk depending on how much she is giving, even small herds needed more than one milker and we used to chat about all sorts of things. Many of the men I worked with were older than me and so they told me about their experiences and passed on a lot of knowledge. Hand milking was the only job that I got really skilled at. I should have done because I estimate that I sat under a cow about 50,000 times during the time I was doing farm work.

Now seems like a good time to mention the second charismatic cow that I met during my time on farms. Her name was Wakey and she was the leader of the Ayrshire herd. Her most impressive attribute was her understanding of what happened to her calves.

Milk produced in winter sold for a higher price and so we calved down most of the cows in September in the fields. As soon as possible after the calf was born we took the Fergie tractor with one of us riding in the link box on the back and collected it.

The older matrons would be gathered round like women drooling over a baby in a pram but we would just shoo them away and they would go back to grazing – including the mother who seemed to forget she had ever had a calf. This meant that no bonding had taken place and there was not the trauma of separation that there would have been if they had been together for any length of time.

Wakey was quite different. Normally we drove away quickly before the mother cottoned on to what was happening but Wakey would come running after us so we drove slowly while she kept her head in the link box to check on the calf.

The cow sheds and calf pens were in the same yard and Wakey would keep a close watch while we installed her calf in a pen. Once she was satisfied she would turn around and walk back to join the herd.

At milking time she would charge into the shed and grab as much food as possible but when the cows were let out after milking she walked over to the pens and mooed to the calves who answered in their treble voices. She was clearly asking them, 'Are my servants looking after you properly?' The calves must have answered in the affirmative because she would turn around and leave with the rest of the herd. I think she looked upon us just as rich people of days gone by looked upon their nannies.

An interesting thing happened to one of the cows on this farm, which I had never seen before. Normally each cow has her place in the herd which she retains but this cow, Susie, lost her position and developed what can only be described as an inferiority complex. For a couple of month after they had dried off, the cows were turned out with the young stock. When they returned to the herd they normally regained their position quickly but poor Susie seemed to have completely lost her confidence. She walked at the back of the herd, even behind the young heifers, and if any of them approached her she would lower her head and put her ears back because she was afraid of being attacked. When the other cows came into the shed she would wait outside until they were tied up and we called her in.

On the farm there were two other workmen besides the cowman. One of these was a middle-aged chap who lived in the only cottage belonging to the farm. He was not as appreciated as he should have been because the farm family didn't have the experience to recognise his skills but he could build a rick better than any man I came across. It started quite small, not much bigger than a table, but would go up to an enormous width so there was no chance of rain driving in. He never had to prop it.

The other chap was the tractor driver called Walter.

He was one of those dangly, bony people with long arms and incredible strength in his fingers. If he turned a tap off you could never get it on again without him. He had absolutely no sense of fear or any idea of the difficulties that he could get into.

There were two tractors on the farm, a blue Fordson and a little grey Fergie. The Fergusons were unique in those days. The implements were attached in such a way that they could be hitched, raised and lowered by hydraulic power. This was a tremendous advantage. With a trailing implement you could not reverse and so a wide headland had to be left, but, if you could lift the plough you could get into a much smaller area.

To get the Ferguson started Walter would hitch it to the back of the Fordson with a short chain and then tow it across the field behind the cowshed until it got going. Once it was running he would have to get out of his tractor and into the Ferguson to stop it before it hit the back of the Fordson. He always got away with it.

Since the land was so steep, tractor work had to be undertaken with great care. It became a bit of a competition between the lads to see who could work a bit further down the bank or plough the extra furrow – but try this too often and you could turn the tractor over. There were no roll bars in those days and how Walter didn't get killed I don't know but we would often find the poor tractor upside-down at the bottom of a bank.

When we got it back on its feet he would say, 'There you are, that's done it some good. It starts beautifully now.'

I heard a story about a couple who had recently bought a farm in Devon and did not know their neighbour very well. One evening they were walking along the top of one of their steep fields when they saw their neighbour on a tractor which had obviously run away with him. He vanished into the bushes at the bottom. They ran down as fast as they could and saw him coming out to meet them. 'Are you all right?' one of them called. 'Well,' he replied, 'I had a bit of a cold last week, but I'm alright now.'

When Walter decided to get married he left for a job where he would be given a cottage and I got the chance to do exactly what I had come to this farm for: look after the tractors. The Wing Commander said, 'Do what it says in the book,' and so I did. I nearly drove everyone mad because I did everything exactly as the book told me it should be done.

Nowadays, ploughs have two sets of bodies and can turn to the right and left so that when you complete a furrow you just swing round on the headland, turn the plough bodies over and start again on the adjacent strip of land. At the time I was learning, the plough was only able to turn furrows to the right so before I started, the field had to be marked out in lands of usually 22 yards wide. I went up one and down the next whilst trying to keep the remaining pieces of unploughed land the same width top and bottom. They never were and I would have to take out bits and pieces here and there to avoid going over the land I had already ploughed which would have led to caustic remarks.

I took great care over setting up the equipment and when we got a new Ferguson with a six-foot reciprocal cutter I followed all the rules for setting it up before I set off. Other people would hitch the thing up and cut an acre before I had started but once I started I could keep going so long as someone came to sharpen the knives for me.

Carrying the silage was great fun. While the others did the milking I would go out and cut enough grass for the day. After it had wilted for a few hours, I put a buckrake on the back of the tractor to collect it. It was something like the old-fashioned hay sweeps with long prongs and I simply drove backwards underneath the cut grass until I had as much as the hydraulic lift could carry. We had dug a pit as deep as we could in the stony ground and I just ran over the top and pulled a lever to drop the grass. We used to have a competition to see who could carry the most loads in a given time. I think the most we ever did was sixteen loads in an hour.

One day I was out buckraking in a thunderstorm and was terrified I would be killed. It had not started raining so the Boss told me to carry on because rain on the cut grass would not have done it a lot of good. The lightning was alarming, and running back empty across the field with metal prongs pointing heavenwards seemed like asking to be struck.

During the same storm, a lifeguard who was standing in bare feet on wet sand decided that it was not safe for anyone to be in the sea. As soon as he put the whistle into his mouth to call them in he was struck dead.

The farm in Bude was the first one that I worked on which didn't have working horses and although I was fond of the Fergie tractor it wasn't quite the same. The tractor did run away once or twice (pushed down a hill by an implement too heavy for the poor little thing) but it never had the sense to turn for home.

One day I watched it chase the Governor round the yard which was very funny but could have ended in tragedy. I backed the tractor into the cow shed doorway to fill the link box with manure and, because the ground sloped, left it in gear. It was started by a handle and the Governor thought he would help by starting it, but since it was in gear it set off after him. He tried to jump out of the way but the wheel hit a large stone and veered towards him. There were several such stones and it looked for all the world as though the Fergie was trying to run him down.

I often talked to the tractor but when you greeted it

in the morning it just sat there whereas most horses would look around and be pleased to see you. Apart from anything else, horses smell nicer. Most animals frankly stink and after you have been in their company for long you no longer realise that you smell too. Horses were different and had a pleasant aroma, unlike the fumes of a tractor.

I missed the horses when I went to work in Bude but they were already too slow and each one or two needed a man to work with them so it made them expensive. I came along as farm horses were on their way out.

I worked for this family for three years and they were the best people who I had ever worked for. They even paid me for the work I did in the house. We did the washing up only when we had used all the crockery and I often tackled this on a Sunday afternoon to fill in the time before milking. They insisted that I put this time down.

I ate dinner with the family and always changed first, not into anything smart, just clean. We would sit and talk because there was very little television in those days and the family hadn't spent money on anything that wasn't vital. There was a wireless that we referred to as 'the box' and to get it to play we had to turn it in the right direction and hit it in the right place.

When the boys were home for the school holidays

we would play card games, table tennis and cricket. On one memorable occasion a tennis court was hired in full view of the public in Bude and the Wing Commander sent a ball whizzing past me at great speed, saying, 'The last time I played tennis was in Monte Carlo.' Well, I never even saw the ball and after a few times I was relegated to playing with the boys.

I was sorry to leave there but it dawned on me that I was never going to be able to raise the capital to have my own farm and in those days it was unheard of for a woman to get a farm manager's job. Reluctantly I decided that I had to give up farm work and go back to school. Of course, I continued to study agriculture.

About the Author

AFTER having spent eleven years on the farms described in this book, Sonia studied at Newcastle-upon-Tyne university where she completed a degree in general agriculture and specialised in agricultural economics. This led to a successful career with the Department of Farm Economics, School of Agriculture, University of Cambridge where she remained for seven years.

Sonia was keen to return to practical farming. She and her husband ran their own market garden for twenty-five years. She lives near Cambridge with her dog, Pippa.